An Introduction Fourier Series and Integrals

Robert T. Seeley

Professor Emeritus
University of Massachusetts at Boston

Dover Publications, Inc.
Mineola, New York

Copyright

Copyright © 1966, 1994 by Robert T. Seeley
All rights reserved.

Bibliographical Note

This Dover edition, first published in 2006, is an unabridged republication of the edition first published by W. A. Benjamin, Inc., New York, in 1966.

International Standard Book Number

ISBN-13: 978-0-486-45307-1
ISBN-10: 0-486-45307-3

Manufactured in the United States by Courier Corporation
45307302
www.doverpublications.com

Editors' Foreword

Mathematics has been expanding in all directions at a fabulous rate during the past half century. New fields have emerged, the diffusion into other disciplines has proceeded apace, and our knowledge of the classical areas has grown ever more profound. At the same time, one of the most striking trends in modern mathematics is the constantly increasing interrelationship between its various branches. Thus the present-day students of mathematics are faced with an immense mountain of material. In addition to the traditional areas of mathematics as presented in the traditional manner—and these presentations do abound—there are the new and often enlightening ways of looking at these traditional areas, and also the vast new areas teeming with potentialities. Much of this new material is scattered indigestibly throughout the research journals, and frequently coherently organized only in the minds or unpublished notes of the working mathematicians. And students desperately need to learn more and more of this material.

This series of brief topical booklets has been conceived as a possible means to tackle and hopefully to alleviate some of these pedagogical problems. They are being written by active research mathematicians, who can look at the latest developments, who can use these developments to clarify and condense the required material, who know what ideas to underscore and what techniques to stress. We hope that these books will also serve to present to the able undergraduate an introduction to contemporary research and problems in mathematics, and that they will be sufficiently informal that the personal tastes and attitudes of the leaders in modern mathematics will shine through clearly to the readers.

Not long after the Calculus became an essential mathematical tool, J. Fourier began his investigations into the study of heat transfer. This research depended upon a special representation of functions in series form; however, at that time such representation, and indeed the general nature of functions, was little understood. One of the major activities of nineteenth-century mathematics centered around Fourier's series. This study played a major role in the founding of modern methods in mathematics, and the topic of Fourier series is still an active and important mathematical field.

This book takes us through the investigations of those nineteenth-century mathematicians, bringing to light the nature of their discoveries and its impact on modern mathematical thought. It brings the reader to Lebesgue's door (with a glimpse inside), giving him the materials necessary to comprehend the importance of Lebesgue's new concept of integration.

Robert Gunning
Hugo Rossi

Princeton, New Jersey
Waltham, Massachusetts
December 1965

Preface

Mathematics as we know it today has developed out of certain inherently interesting problems, among the oldest of which are those connected with accounting, surveying, and astronomy. By the early nineteenth century, problems such as the vibration of strings and the conduction of heat were analyzed with a good deal of success, and one of the main tools that made this possible was the use of infinite series of sines and cosines. (The first to apply such series extensively was Joseph Fourier, and hence they bear his name.) But the infinite has always posed serious problems for our intuition, and the growing importance of infinite series caused a number of investigators to concentrate on the logical and philosophical questions involved. This tendency toward specialization has gone so far today that most students in engineering and physics learn Fourier's results with very little attention to the logical questions, while students in mathematics learn a list of theorems whose origins are only suggested in an occasional footnote or exercise.

This book attempts, by its organization, to show the interplay of physics and mathematics, which is perhaps the most interesting aspect of the subject. Beginning with a physical problem, it sets up and analyzes the mathematical model, establishes the principal properties, and proceeds to apply these results and methods to new situations. The final chapter, on Fourier transforms, derives analogs of the results obtained for Fourier series, and finally reapplies them in the analysis of a problem of heat conduction.

The writing of these notes began in 1960 at Harvey Mudd College, where the first chapter was used to supplement the course in advanced calculus. The development into a book came in the summer of 1965. I appreciate very much the thoughtful comments and encouragement of Jay Martin Anderson, Robin Ives, Lorraine Liggins, Wilhelm Magnus, Kenneth Miller, Hugo Rossi, Silvan Schweber, and my wife, Charlotte.

Robert T. Seeley

Waltham, Massachusetts
January 1966

Contents

Introduction

Reading a book is something like taking a guided journey. This introduction describes for the reader primarily what preparation he should bring with him, with some recommendations as to what is expected along the way, and what shortcuts are possible.

The chapters have been arranged so that the reader who stops at the end of any one of them will have a coherent picture of some part of the subject. This does not imply that the entire book gives a complete picture, for we present only a few topics selected for their combined merits of accessibility and significance. It is possible to go directly from Chapter 1 to Chapter 3, but no other skipping is recommended, and Chapter 1 in particular is fundamental to the whole book. All the exercises are to be read, even if they are not worked out, since later parts of the text occasionally rely on them. Some provide simple applications of the text material; some test the reader's understanding more deeply by applying the ideas, rather than the explicitly formulated results, to establish new results. A few exercises even require ideas not suggested by the text.

The preparation assumed on the part of the reader has been kept to the minimum necessary to present actual proofs, and to cover the most important examples. The following paragraphs list the notations and principal results used. The reader who understands the proofs of these results should be prepared to follow the arguments presented in this text.

A set in the plane is sometimes denoted by putting its description in brackets; e.g., the disk of radius one is written as

$$\{(x, y): x^2 + y^2 < 1\}$$

and the graph of a real-valued function f of one variable is

$$\{(x, y): y = f(x)\}.$$

We also consider complex-valued functions of one or two real variables, in particular the function

$$e^{i\theta} = \cos \theta + i \sin \theta.$$

A function f defined on an interval $a < x < b$ is called *piecewise continuous* if there are finitely many points a_j, $a = a_0 < a_1 < \cdots < a_n = b$, such that f is continuous in the intervals $a_j < x < a_{j+1}$, and the one-sided limits $f(a_j+)$ exist for $0 \leq j < n$, and $f(a_j-)$ exists for $0 < j \leq n$. By $f(a_j+)$ we mean the limit from the right,

$$\lim_{x \to a_j, x > a_j} f(x)$$

and by $f(a_j-)$ the limit from the left. Piecewise continuous functions have graphs like that in Fig. 0-1. If in addition the derivative f' is continuous in each of the intervals $a_j < x < a_{j+1}$, and the limits $f'(a_j+)$ and $f'(a_j-)$ exist, then f is called *piecewise differentiable*. We will consider only piecewise continuous functions, but the reader familiar with Riemann integrable functions, or even Lebesgue integrable functions, may in most cases substitute them for the piecewise continuous ones without arriving at a false theorem.

(Actually, the consideration by mathematicians of functions with discontinuities is due largely to the fact that certain series of sines and cosines have limits of this type. It was in connection with the study of Fourier series that the concept of

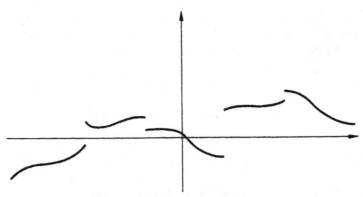

FIGURE 0-1. *Graph of piecewise continuous function.*

function was enlarged from "something given by an explicit formula," like

$$f(x) = \cos(e^{\sin x})$$

to "something which assigns a number to each member of some set of real numbers" in any way, no matter how capricious, like

$$g(x) = 1 \quad \text{if} \quad x \text{ is rational}$$

$$g(x) = 0 \quad \text{if} \quad x \text{ is irrational.})$$

The main operations on functions of one variable will be integration and summation of series. We use the following notations for the integral of f from a to b:

$$\int_a^b f \qquad \int_a^b f(x)\, dx \qquad \int_a^b f(t)\, dt \qquad \text{etc.}$$

The *fundamental theorem of calculus* is two results. One states that, if f is continuous, then the function

$$F(x) = \int_a^x f$$

has as derivative the function f: $F'(x) = f(x)$. The second states that, if F has the continuous derivative f, then

$$F(b) - F(a) = \int_a^b f. \tag{0-1}$$

This result extends immediately to the case that F is continuous and piecewise differentiable; then the values selected for f at the points where F' fails to exist do not affect the integral in (0-1). If F and G are both continuous and piecewise differentiable, then the formula for integration by parts is valid:

$$\int_a^b FG' = F(b)G(b) - F(a)G(a) - \int_a^b F'G.$$

When f is any bounded real-valued function defined on an interval $a < x < b$, then $\sup f$, or $\sup_{a<x<b} f(x)$, denotes the least upper bound of all values assumed by f on the interval. If f is continuous on a closed, bounded interval $a \leq x \leq b$, then it is bounded and $\sup_{a \leq x \leq b} f(x)$ is actually one of the values assumed by f; in this case it may also be denoted by $\max f$. The corresponding notations for lower bounds are *inf* and *min*.

About ordinary differential equations we assume it known that equations of the form

$$y'' + ay' + by = 0$$

with constant coefficients a and b are solved by setting $y = e^{rx}$ and solving for r; and that "equidimensional" or Cauchy equations like

$$x^2 y'' + axy' + by = 0$$

are solved by setting $y = x^r$ and solving for r. If the resulting equation for r has a double root $r = r_0$, then there is also a solution $xe^{r_0 x}$ in the constant coefficient case or $x^{r_0} \log x$ in the equidimensional case.

In general, the functions of two variables that we consider are continuous. If they have partial derivatives, they are denoted by $\partial f / \partial x$, $\partial^2 f / \partial x \partial y$, $\partial^2 f / \partial x^2$, and so on. At times it is convenient to use subscripts to denote partial derivatives, particularly when they must be evaluated at a particular point. Thus $f_x(x_0, y_0)$ is the value of the partial derivative $\partial f / \partial x$ at the point (x_0, y_0). The context should make it clear when subscripts do *not* indicate partial derivatives.

If f is a continuous function of x and y, then

$$\int_a^b f(x, y) \, dx \tag{0-2}$$

is a continuous function of y and

$$\int_c^d \left[\int_a^b f(x, y) \, dx \right] dy = \int_a^b \left[\int_c^d f(x, y) \, dy \right] dx.$$

If $\partial f / \partial y$ is continuous, then the function given by the integral in (0-2) has a derivative with respect to y given by

$$\int_a^b f_y(x, y) \, dx.$$

This is Leibnitz's rule for differentiation under the integral.

Last, but not least, some knowledge of infinite series is required. For instance, we invoke the convergence of alternating series, and the validity of term-by-term integration and differentiation of series of functions under appropriate conditions, as well as the continuity of the sum of a uniformly convergent series of continuous functions.

The text contains several numerical reference systems. Section 1-1 is Section 1 of Chapter 1; Theorem 1-1 and Exercise 1-1 are Theorem 1 and Exercise 1 of Chapter 1; and (1-1) refers to the first numbered display in Chapter 1. A number in brackets refers to the Bibliography at the end of the text.

With these preliminaries out of the way, we hope the remainder of our book will offer the reader not only some instruction, but also some of the pleasures of understanding and discovery.

1

Dirichlet's Problem and Poisson's Theorem

This entire chapter discusses a single problem that lies at the heart of the subject of Fourier series. In physical terms, it is to determine the steady state temperature distribution in a disk when the boundary temperatures are known; the mathematical formulation is known as *Dirichlet's problem*. Starting from scratch, we arrive at a solution in the form of a series $\Sigma\, a_n r^{|n|} e^{in\theta}$, called a *trigonometric series*. The attempt to verify that this can actually be made to solve the heat problem leads to Poisson's theorem, one of the most important and attractive elementary theorems of analysis. A number of important consequences are evolved below, but these few only suggest the importance of Poisson's theorem. Echoes of the proof, and of the integral representation on which it is based, appear frequently in mathematics even today.

1-1. THE EQUATION OF STEADY STATE HEAT CONDUCTION

The first step in solving heat problems (as in most problems of mathematical physics) is to find a differential equation governing the situation. Since we are concerned with a disk, the natural coordinates are polar. The temperature at the point with coordinates (r, θ) is denoted by $u(r, \theta)$.

In order to find the relevant equation, consider any section of the disk given by

$$0 < r_0 < r < r_1 \leq 1 \qquad \theta_0 < \theta < \theta_1 \qquad (1\text{-}1)$$

(see Fig. 1-1). Since we are considering a steady state, the rate at which heat flows into this section must be 0; otherwise the average temperature would change with time. Now it is a basic postulate of heat conduction that the rate at which heat crosses a curve C is proportional to the integral along C of the normal derivative $\partial u / \partial n$ of the temperature distribution. Here $\partial u / \partial n$ is the derivative of u with respect to arc length along any curve perpendicular to C. When C is the side $\theta = \theta_1$ of the portion given in (1-1), we can take these perpendicular curves to be given by $r =$ constant. Then, since the length of a circular arc is the angle times the radius, along

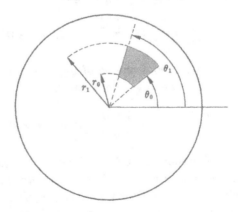

FIGURE 1-1. *Derivation of the heat equation in polar coordinates.*

$\theta = \theta_1$ the normal derivative is

$$\frac{du}{dn} = \lim_{h \to 0} \frac{u(r, \theta_1 + h) - u(r, \theta_1)}{rh}$$

$$= r^{-1}u_\theta(r, \theta_1)$$

and the rate at which heat flows into the section in Fig. 1-1 along the boundary $\theta = \theta_1$ is

$$k \int_{r_0}^{r_1} r^{-1}u_\theta(r, \theta_1) \, dr,$$

where k is the conductivity. Adding corresponding expressions for the other three boundaries and setting the net flow equal to zero, we get

$$\int_{r_0}^{r_1} r^{-1}[u_\theta(r, \theta_1) - u_\theta(r, \theta_0)] \, dr$$

$$+ \int_{\theta_0}^{\theta_1} [r_1 u_r(r_1, \theta) - r_0 u_r(r_0, \theta)] \, d\theta$$

$$= 0.$$

If we divide by $\theta_1 - \theta_0$ and let $\theta_1 \to \theta_0$, this yields

$$\int_{r_0}^{r_1} r^{-1}u_{\theta\theta}(r, \theta_0) \, dr + r_1 u_r(r_1, \theta_0) - r_0 u_r(r_0, \theta_0) = 0.$$

(The first term comes from Leibnitz's rule, and the second from the fundamental theorem of calculus.) Now, dividing by $r_1 - r_0$ and letting $r_1 \to r_0$, we get

$$\frac{1}{r_0} u_{\theta\theta}(r_0, \theta_0) + (r u_r)_r(r_0, \theta_0) = 0.$$

Since r_0, θ_0 are the coordinates of any point inside the disk, except the center, we have the equation

$$\frac{\partial(r \, \partial u/\partial r)}{\partial r} + \frac{1}{r}\frac{\partial^2 u}{\partial \theta^2} = 0 \qquad (r \neq 0). \tag{1-2}$$

This is the polar coordinate form of *Laplace's equation*. Solutions of this equation are called *harmonic functions*.

The above derivation does not prove anything about *actual* temperature distributions. It starts with some (reasonable) assumptions about the connection between heat flow and the derivatives of the temperature distribution, and arrives mathematically at Eq. (1-2). The steps leading from these assumptions are valid from the point of view of mathematics, but it cannot be proved that they are justified physically. However, in this case it turns out (experimentally) that intuition has not led us astray, and the connection between Eq. (1-2) and our conception of heat flow provides a large supply of questions about Eq. (1-2). We shall not here consider in much detail the other aspect of this connection, that is, how the answers to these questions enter into the understanding of physics, and even into the design of structures where heat plays a role.

> **Exercise.** 1-1. Derive the rectangular coordinate form of the steady state heat equation.

Because of the peculiarity of polar coordinates at $r = 0$, Eq. (1-2) does not apply there, and the question arises whether some condition must be applied at that point. Since u is required to have continuous first partials at all other points (implicit in writing the derivatives in (1-2)), a natural condition might be to require that $\partial u/\partial x$ and $\partial u/\partial y$ be continuous at the origin. However, for most of our discussion we can develop things nicely by requiring only that

$$u(r, \theta) \quad \text{be continuous for} \quad 0 \le r \le 1. \qquad (1\text{-}3)$$

This condition has the advantage that it can be stated easily in polar coordinates, and is simple to check. It is obviously a necessary condition, from the physical point of view.

Another peculiarity of polar coordinates is that $(r, \theta + 2\pi)$ denotes the same point as (r, θ). Accordingly we require

$$u(r, \theta + 2\pi) = u(r, \theta). \qquad (1\text{-}4)$$

It is clear from the physical point of view that (1-2), (1-3), and (1-4) alone cannot determine a solution of the heat problem uniquely; for instance, *any* constant u satisfies them. We complement them by specifying the temperature distribution on the boundary $r = 1$:

$$u(1, \theta) = f(\theta). \tag{1-5}$$

Again, $f(\theta + 2\pi) = f(\theta)$.

The problem of finding a function u satisfying Eqs. (1-2) to (1-5) for a given f is called a *Dirichlet problem*. This phrase covers a large variety of problems with certain features in common with the one before us now, but specifically in this text it means the problem of finding a solution u of Laplace's equation in some region of the plane, when the values of u are prescribed on the boundary of the region.

> ***Exercise.*** **1-2.** Which of the conditions (1-2)–(1-4) does $u = \log r$ satisfy? Find a function u satisfying (1-2) and (1-3), but not (1-4).

1-2. THE SOLUTION OF THE DIFFERENTIAL EQUATION BY PRODUCTS

In trying to satisfy (1-2)–(1-5) we look first for solutions of (1-2). The technique used, called the *separation method*, reduces the partial differential equation (1-2) to familiar *ordinary* differential equations by restricting attention temporarily to functions u of the special form

$$u(r, \theta) = R(r)\Theta(\theta).$$

The rationale for this is that the partial derivatives of such a function u involve ordinary derivatives of R and Θ. For $u = R\Theta$, (1-2) becomes

$$\frac{\partial}{\partial r}(rR'\Theta) + \frac{1}{r}R\Theta'' = 0$$

or

$$r^2R''\Theta + rR'\Theta + R\Theta'' = 0.$$

This mixture of two ordinary differential equations can be sorted out if we divide by $R\Theta$, obtaining

$$\frac{r^2R''}{R} + \frac{rR'}{R} = -\frac{\Theta''}{\Theta}.$$

Now the left side of this equation is, by its form, independent of θ, while the right side is independent of r. Since they are equal, they are both independent of r and θ, that is, constant:

$$\frac{r^2 R''}{R} + \frac{rR'}{R} = -\frac{\Theta''}{\Theta} = c.$$

This yields related ordinary differential equations for R and Θ. Consider first the equation for Θ, $\Theta'' + c\Theta = 0$. The form of the general solution of this equation depends on the sign of c; it is easy to check that it is given by

$$
\begin{aligned}
\Theta(\theta) &= A \exp(i\sqrt{c}\,\theta) + B \exp(-i\sqrt{c}\,\theta) && \text{if } c > 0 \\
&= A + B\theta && \text{if } c = 0 \\
&= A \exp(\sqrt{-c}\,\theta) + B \exp(-\sqrt{-c}\,\theta) && \text{if } c < 0.
\end{aligned}
$$

Similarly the equation in R, $r^2 R'' + rR' - cR = 0$, has the solution

$$
\begin{aligned}
R(r) &= ar^{\sqrt{c}} + br^{-\sqrt{c}} && \text{if } c > 0 \\
&= a + b \log r && \text{if } c = 0 \\
&= ar^{i\sqrt{-c}} + br^{-i\sqrt{-c}} && \text{if } c < 0.
\end{aligned}
$$

Thus, for instance, $(A + B\theta)(a + b \log r)$ satisfies the equation (1-2), and generally the product of any $\Theta(\theta)$ and $R(r)$ *with the same value of c* yields a solution of (1-2). We are also interested in conditions (1-3) and (1-4). Now the solution for Θ when $c < 0$ is unbounded unless $A = B = 0$, and hence does not yield any nonzero solution satisfying (1-4); for a continuous function Θ of period 2π is bounded by its maximum on the interval $0 \leq \theta \leq 2\pi$. Similarly, in the case $c = 0$, we must take $B = 0$ to satisfy (1-4). The solution for Θ when $c > 0$ has period 2π precisely when $\sqrt{c}$ is an integer, or $c = n^2 \geq 0$. Thus applying (1-4) restricts us to the possibilities $c = n^2$, $n = 0, 1, 2, \ldots$. Applying condition (1-3) to the solution for R with these values of c, we find that $b = 0$ in all cases. Now the remaining product solutions

$$
\begin{aligned}
u_n(r, \theta) &= Aa && (n = 0) \quad \text{or} \\
&= ar^n(Ae^{in\theta} + Be^{-in\theta}) && (n > 0)
\end{aligned}
$$

are the *only* product solutions $u = R\Theta$ satisfying (1-2), (1-3), and (1-4).

This is a fair collection, but it is clear that none of these solutions has enough flexibility to satisfy a very wide range of choices for the function f in (1-5). However, it is easy to check that if $u_1, \ldots, u_N$ are solutions of (1-2), (1-3), and (1-4), then so is $u = u_1 + \cdots + u_N$. Thus

$$A_0 + \sum_1^N r^n(A_n e^{in\theta} + B_n e^{-in\theta})$$

is also a solution of (1-2), (1-3), and (1-4). Changing the names of the arbitrary constants, we can write this more simply as $\sum_{-N}^N A_n r^{|n|} e^{in\theta}$. On the boundary it assumes the value

$$u(1, \theta) = \sum_{-N}^N A_n e^{in\theta}.$$

Since there is no limit to the size of N here, we are tempted to try

$$u(r, \theta) = \sum_{-\infty}^{\infty} A_n r^{|n|} e^{in\theta} \tag{1-6}$$

which has, at least from a formal point of view, the boundary values

$$u(1, \theta) = \sum_{-\infty}^{\infty} A_n e^{in\theta}. \tag{1-7}$$

Now, we have as yet no logical reason to think this is a "general" solution; but it turns out to be just that. What Fourier asserted, and we shall eventually show, is that the coefficients in (1-7) can be chosen so that (1-7) represents virtually any function f that we choose. Using these same coefficients in (1-6) then provides a solution of the temperature problem originally posed, namely a function u satisfying (1-2)–(1-5).

The question of how to represent f by the series (1-7) is considered in the next section, and then the proof that the solution so obtained satisfies (1-2)–(1-5) is taken up.

1-3. *THE FOURIER COEFFICIENTS*

We now turn to the question: if $f(\theta)$ is a given function of period 2π, can the coefficients A_n in (1-7) be chosen so that $u(1, \theta) = f(\theta)$, that is, so that

$$f(\theta) = \sum_{-\infty}^{\infty} A_n e^{in\theta} \qquad (1\text{-}8)$$

and if so, how? Actually, it is more reasonable to see first how the coefficients should be chosen, then investigate whether choosing them this way really works; we answer the first question in this section.

As an indication of how to begin, recall the determination of the coefficients a_n in the Taylor series $g(x) = a_0 + a_1 x + a_2 x^2 + \cdots$. It is clear from a glance at the series that a_0 is $g(0)$; and it is not too hard to see that the way to get at a_1 is to differentiate once and set $x = 0$, obtaining $a_1 = g'(0)$, and so to find generally that $n! a_n = g^{(n)}(0)$. Turning to the series (1-8) of exponentials, it is not quite so obvious what the precise significance of A_0 is; certainly it cannot be found by substituting any particular value of θ, like $\theta = 0$. However, observe that the average value of $e^{in\theta}$ over any interval of length 2π is zero, except when $n = 0$. Thus A_0 should be the average value of the right side of (1-8), so that

$$A_0 = \frac{1}{2\pi} \int_{-\pi}^{\pi} f(\theta) \, d\theta.$$

This suggests that the formulas for the other coefficients will involve integrals of f. In fact, for any N we can make A_N the only coefficient multiplying a term of nonzero average value by writing

$$f(\theta)e^{-iN\theta} = \sum_{-\infty}^{\infty} A_n e^{i(n-N)\theta},$$

so that in general we expect to have

$$A_N = \frac{1}{2\pi} \int_{-\pi}^{\pi} f(\theta)e^{-iN\theta} \, d\theta. \qquad (1\text{-}9)$$

In deducing these formulas we have essentially relied on term-by-term integration of the series (1-8), which may or may not be valid. However, the procedure seems highly likely, so likely that we *take the formula (1-9)* as the *definition of the Fourier coefficients of f.* The series $\Sigma_{-\infty}^{\infty} A_n e^{in\theta}$ with A_n given by (1-9) is called the *Fourier series of f.*

One thing should be emphasized; as yet we do *not* know that for any arbitrary f, f equals its Fourier series. What we do have is some justification for thinking that if f equals any such series, the coefficients should be determined by (1-9).

For the heat problem originally proposed, then, we are led to the *likelihood* that if we take $u(r, \theta)$ to be the series (1-6), with coefficients given by (1-9), then $u(r, \theta)$ will satisfy (1-2)–(1-5).

Exercises. **1-3.** Find the Fourier series of $f_1(\theta) = \sin^2 \theta$; of $f_2(\theta) = \cos^3 \theta$.

1-4. Suppose $f(\theta) = |\theta|$ for $-\pi < \theta < \pi$. Find the formal series solution of the corresponding heat problem in the disk. How many terms of the series will give $u(r, \theta)$ with an error < 0.1 throughout the disk? Evaluate $u(\frac{1}{2}, \pi)$ to two decimals. Show that $u(r, \pm\pi/2) = \pi/2$.

1-5. Supposing f is bounded and A_n is given by (1-8), prove that $\sup_n |A_n|$ is finite.

1-6. How should the coefficients in (1-6) be chosen so that $u_r(1, \theta) = g(\theta)$ for a prescribed g? When u is given by (1-6), what is the average of $u_r(1, \theta)$ over $-\pi < \theta < \pi$? Can the coefficients in (1-6) be chosen so that $u_r(1, \theta) = 1$? (These questions relate to the *Neumann problem*, of finding the steady temperature distribution when the rate of heat loss along the boundary is given.)

1-4. POISSON'S KERNEL

Substituting in (1-6) the values of A_n given by (1-9), we get for the formal Fourier series solution of (1-2)–(1-5)

$$u(r, \theta) = \sum_{-\infty}^{\infty} r^{|n|} e^{in\theta} \frac{1}{2\pi} \int_{-\pi}^{\pi} f(t) e^{-int} \, dt$$

$$= \sum_{-\infty}^{\infty} \int_{-\pi}^{\pi} \frac{1}{2\pi} r^{|n|} e^{in(\theta - t)} f(t) \, dt.$$

In attempting to find out to what extent this expression actually satisfies the problem posed in Section 1-1, we simplify it as far as possible. For $r < 1$, the series $\sum_{-\infty}^{\infty} r^{|n|} e^{in(\theta-t)}$ converges uniformly in t, so we may integrate it term by term to obtain

$$u(r,\theta) = \frac{1}{2\pi} \int_{-\pi}^{\pi} \sum_{-\infty}^{\infty} r^{|n|} e^{in(\theta-t)} f(t) \, dt.$$

Again for $r < 1$, we have from the familiar geometric series that

$$\sum_{0}^{\infty} r^n e^{in(\theta-t)} = \frac{1}{1 - re^{i(\theta-t)}}$$

and

$$\sum_{-\infty}^{-1} r^{-n} e^{in(\theta-t)} = \frac{re^{-i(\theta-t)}}{1 - re^{-i(\theta-t)}}.$$

Adding these yields

$$u(r,\theta) = \frac{1}{2\pi} \int_{-\pi}^{\pi} \frac{1 - r^2}{1 - 2r\cos(\theta-t) + r^2} f(t) \, dt \qquad (r < 1). \tag{1-10}$$

This is called *Poisson's integral*. The function

$$P(r,\phi) = \frac{1}{2\pi} \frac{1 - r^2}{1 - 2r\cos\phi + r^2} \tag{1-11}$$

is called *Poisson's kernel*. Formula (1-10) gives the formal solution u in a very neat and useful form, at least from the theoretical point of view. In the next section we use this form to investigate whether the formal solution satisfies requirement (1-5); the verification of the other conditions is left as an exercise.

Exercises. 1-7. Check that the expression (1-10) satisfies (1-2)–(1-4); if necessary, assume f to be continuous.

1-8. Prove that $r^n e^{in\theta}$ and $r^n e^{-in\theta}$ are polynomials in x and y, where $n \geq 0$, $x = r\cos\theta$, $y = r\sin\theta$. (These are the polynomial solutions of degree n of Laplace's equation, and are called harmonic polynomials.)

1-5. *ASSUMPTION OF BOUNDARY VALUES:*
POISSON'S THEOREM

Before proceeding to the present results, the reader should be warned that we are not going to prove the equality (1-7) quite as it stands; in fact (1-7) is not always true. The difficulty is that all our series operations (primarily term-by-term integration and differentiation) were valid only when $r < 1$; in fact, we do not know if the series for $u(r, \theta)$ even converges when $r = 1$. The integral formula (1-10), too, takes a rather disconcerting form when $r = 1$. However, we *can* show that $u(r, \theta)$ converges to $f(\theta)$ as $r \to 1$, when $u(r, \theta)$ is given by (1-10). From the point of view of the heat problem, this is more relevant than deciding the convergence of (1-7). What we shall actually prove is the following.

Theorem 1-1. *If f is continuous, of period 2π, and $u(r, \theta)$ is given by (1-10), then $\lim_{r \to 1} u(r, \theta) = f(\theta)$, uniformly in θ.*

We refer to this as *Poisson's theorem.* Together with the result of Exercise 1-7, it shows that Poisson's integral (1-10) provides a solution of the problem posed by Eqs. (1-2) to (1-5). In order to understand why

$$\lim_{r \to 1} \int_{-\pi}^{\pi} P(r, \theta - t) f(t) \, dt = f(\theta)$$

it is very helpful to sketch $P(r, \phi)$ as a function of ϕ, for various values of r, for instance $r = 0, r = \frac{1}{2}, r = \frac{3}{4}$, and so on. We list some of the salient features of the graph of $P(r, \phi)$ as a function of ϕ.

(i) $P(r, \phi)$ is even in ϕ, that is, $P(r, -\phi) = P(r, \phi)$.

(ii) The maximum of $P(r, \phi)$ occurs at $\phi = 0$, and is $(1 + r)/2\pi(1 - r)$.

(iii) $P(r, \phi)$ is monotone decreasing in $0 \leq \phi \leq \pi$, achieving a minimum of $(1 - r)/2\pi(1 + r)$ at $\phi = \pi$; thus $P > 0$.

(iv) $\int_{-\pi}^{\pi} P(r, \phi) \, d\phi = 1$. (This is easily checked by integrating the series for $P(r, \phi)$.)

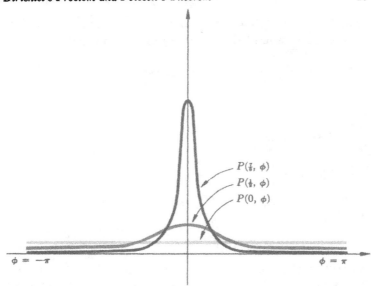

FIGURE 1-2. *Graphs of $P(r, \phi)$ for $r = 0$, $r = \frac{1}{2}$, $r = \frac{7}{8}$.*

Sketches are given in Fig. 1-2 for $P(0, \phi)$, $P(\frac{1}{2}, \phi)$, and $P(\frac{7}{8}, \phi)$. Now suppose $f(t)$ has the graph sketched in Fig. 1-3, and we are evaluating $u(r, 0)$. According to (1-10), we do so by multiplying by $P(r, \theta - t)$, which has its peak at $t = \theta$, then integrating from $-\pi$ to π. Actually, since both P and f have period 2π in the variable t, we could just as well integrate

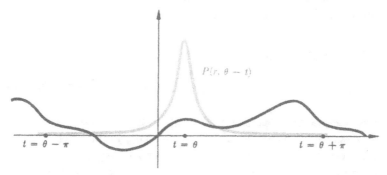

FIGURE 1-3. *Illustration of Poisson integral $\int_{-\pi}^{\pi} P(r, \theta - t) f(t) \, dt$.*

from $\theta - \pi$ to $\theta + \pi$; either way can be thought of as integrating once around the boundary of the unit disk.

Exercise. 1-9. Suppose $g(t)$ has period 2π. Prove

$$\int_{a-\pi}^{a+\pi} g(t) \, dt = \int_{-\pi}^{\pi} g(t) \, dt$$

for any number a.

The integral

$$\begin{aligned} u(r, \theta) &= \int_{-\pi}^{\pi} P(r, \theta - t) f(t) \, dt \\ &= \int_{\theta-\pi}^{\theta+\pi} P(r, \theta - t) f(t) \, dt \end{aligned}$$

can be thought of as a "weighted" average of f. If $r = 0$, then $P = 1/2\pi$, and $u(0, \theta)$ is the usual (that is, the unweighted) average of the boundary temperatures. As r increases while θ remains fixed, the values of $f(t)$ in the vicinity of $t = \theta$ are weighted more heavily, and the others less heavily. This certainly seems reasonable from the physics of the situation; the nearby boundary temperatures have the greatest effect on u. It also suggests why $u(r, \theta)$ approaches $f(\theta)$ as $r \to 1$. The proof of Theorem 1-1 is based on this view of the relation between f and u.

Proof of Theorem 1-1. Given $\epsilon > 0$, we must find $r_0 < 1$ so that, for all r in $r_0 < r < 1$, $|u(r, \theta) - f(\theta)| < \epsilon$ holds. By property (iv) of the Poisson kernel,

$$\begin{aligned} |u(r, \theta) - f(\theta)| &= \left| \int_{\theta-\pi}^{\theta+\pi} P(r, \theta - t)[f(t) - f(\theta)] \, dt \right| \\ &\leq \int_{\theta-\pi}^{\theta+\pi} P(r, \theta - t) |f(t) - f(\theta)| \, dt. \end{aligned}$$

The inequality uses the fact that $P \geq 0$. Now since f is continuous, we can make $|f(t) - f(\theta)|$ small when t is near θ; and from the nature of $P(r, \theta - t)$, we can make P small by taking r close to 1, when t is *not* near θ. This suggests breaking the integral into two parts, and using the two effects separately. We have for any number δ, with $0 < \delta < \pi$, that

$$|u(r,\theta) - f(\theta)| \leq \int_{\theta-\pi}^{\theta+\pi} P(r, \theta - t)|f(t) - f(\theta)|\, dt$$

$$= \int_{|\theta-t|\leq\delta} + \int_{\delta<|\theta-t|<\pi} = I_1 + I_2,$$

where I_1 is the integral for $|\theta - t| \leq \delta$, and I_2 the integral for $\delta < |\theta - t| < \pi$. Since f is continuous on $-2\pi \leq t \leq 2\pi$, it is uniformly continuous there, and we can choose the above δ so that $|f(t) - f(\theta)| < \epsilon/2$ whenever $|\theta - t| \leq \delta$. Then

$$I_1 \leq \int_{|t-\theta|\leq\delta} P(r, \theta - t)\frac{\epsilon}{2}\, dt < \frac{\epsilon}{2}\int_{\theta-\pi}^{\theta+\pi} P(r, \theta - t)\, dt = \frac{\epsilon}{2}.$$

Considering the remaining term,

$$I_2 < \frac{1}{2\pi}\frac{1 - r^2}{1 - 2r\cos\delta + r^2}\int_{\delta<|t-\theta|<\pi} |f(t) - f(\theta)|\, dt$$

since

$$\max_{\delta<|t-\theta|<\pi} P(r, \theta - t) = \frac{1}{2\pi}\frac{1 - r^2}{1 - 2r\cos\delta + r^2}.$$

Now for a fixed δ between 0 and π,

$$\lim_{r\to1}\frac{1}{2\pi}\frac{1 - r^2}{1 - 2r\cos\delta + r^2}\int_{-\pi}^{\pi} |f(t) - f(\theta)|\, dt = 0$$

so that there is an $r_0 < 1$ such that when $r_0 < r < 1$ this expression is less than $\epsilon/2$, making $I_2 < \epsilon/2$. Then finally for $r_0 < r < 1$,

$$|u(r,\theta) - f(\theta)| \leq I_1 + I_2 < \frac{\epsilon}{2} + \frac{\epsilon}{2} = \epsilon.$$

This proves Theorem 1-1.

Exercises. **1-10.** Prove that $\min_{-\pi\leq\theta\leq\pi} f(\theta) \leq u(r, t) \leq \max_{-\pi\leq\theta\leq\pi} f(\theta)$, for $0 \leq r < 1$, when f is real-valued.

1-11. Suppose that f has period 2π and is piecewise continuous. Prove that

$$\lim_{r\to1-}\int_{-\pi}^{\pi} P(r, \theta_0 - t)f(t)\, dt = f(\theta_0)$$

if f is continuous at $\theta = \theta_0$.

1-12. Suppose f has period 2π, and is continuous on the interval $a < t < b < a + 2\pi$. Prove that, for every $\eta > 0$, $u(r, \theta)$ converges uniformly to $f(\theta)$ on $a + \eta \leq \theta \leq b - \eta$.

1-6. TWO SIMPLE CONSEQUENCES OF POISSON'S THEOREM

It might seem that the previous section settles the Dirichlet problem posed by (1-2)–(1-5), since it provides a neat solution in the form of a relatively simple integral. But there remains the question: are there any *other* solutions? This is considered in Section 1-7 but the discussion there is easier when we know the following important mathematical consequences of Poisson's theorem.

Theorem 1-2. *If f and g are continuous functions of period 2π, and if they have the same Fourier coefficients, then $f = g$.*

Proof. By assumption we have

$$\int_{-\pi}^{\pi} f(\theta)e^{-in\theta} \, d\theta = \int_{-\pi}^{\pi} g(\theta)e^{-in\theta} \, d\theta = a_n.$$

Then by Theorem 1-1, the function

$$u(r, \theta) = \sum_{-\infty}^{\infty} r^{|n|} a_n e^{in\theta}$$

converges to $f(\theta)$, and to $g(\theta)$, as $r \to 1$. Thus $f(\theta) = g(\theta)$.

Exercise. **1-13.** Suppose f and g are piecewise continuous on $[-\pi, \pi]$, and have the same Fourier coefficients. Then $f(\theta) = g(\theta)$ if f and g are both continuous at the point θ. (See Exercise 1-11.)

The following corollary of Theorem 1-2 is a preliminary result giving conditions under which the Fourier series of a function f actually converges to f.

Corollary. *If f is continuous of period 2π, and if the Fourier series of f converges uniformly to some function g, then $g = f$.*

Proof. Let a_n be the Fourier coefficients of f, so $g = \sum_{-\infty}^{\infty} a_n e^{in\theta}$. Since the series is uniformly convergent by assumption, we can calculate the Fourier coefficients b_m of g by

$$b_m = \frac{1}{2\pi} \int_{-\pi}^{\pi} e^{-im\theta} \sum_{-\infty}^{\infty} a_n e^{in\theta} \, d\theta = \frac{1}{2\pi} \sum_{-\infty}^{\infty} a_n \int_{-\pi}^{\pi} e^{i(n-m)\theta} \, d\theta$$

$$= a_m.$$

The function g is the sum of a uniformly convergent series; so it is continuous, and we can apply Theorem 1-2 to find $g = f$.

Exercise. 1-14. Show that the Fourier series for the function $f(\theta) = |\theta|$ ($-\pi \le \theta \le \pi$) converges uniformly to f. Evaluate the particular series obtained for $\theta = 0$.

1-7. UNIQUENESS OF THE SOLUTION OF THE HEAT PROBLEM

As noted in the beginning of the previous section, it is conceivable that there might be other solutions of (1-2)–(1-5); and it is important to know whether or not there are, for then we would have to prescribe more conditions on u in order to determine the temperature distribution. One of the easiest ways to show that conditions (1-2)–(1-5) really determine u is by using Green's theorem, but it is more appropriate in the present context to rely on Fourier series entirely. By going through part of the derivation of the Poisson kernel solution from a new viewpoint and applying Theorem 1-2, we get the following result, called a *uniqueness theorem*.

Theorem 1-3. *If $f(\theta)$ is periodic and continuous, and if $v(r, \theta)$ satisfies conditions (1-2)–(1-4) and $v(r, \theta) \to f(\theta)$ uniformly as $r \to 1-$, then $v(r, \theta)$ is identical with the solution given by the Poisson kernel.*

Proof. For each fixed $r < 1$, $v(r, \theta)$ has a Fourier series $\sum_{-\infty}^{\infty} a_n(r)e^{in\theta}$, with

$$a_n(r) = \frac{1}{2\pi} \int_{-\pi}^{\pi} v(r, \theta)e^{-in\theta} \, d\theta.$$

Taking into account that $v(r, \theta)$ satisfies the equation (1-2), we find by differentiating under the integral that

$$(ra_n')' = \frac{1}{2\pi} \int_{-\pi}^{\pi} \frac{\partial}{\partial r} (rv_r(r, \theta))e^{-in\theta} d\theta$$

$$= \frac{-1}{2\pi} \int_{-\pi}^{\pi} \frac{1}{r} v_{\theta\theta}e^{-in\theta} d\theta.$$

If we integrate by parts, remembering that $v(r, \theta - \pi) = v(r, \theta + \pi)$, we find further

$$(ra_n')' = \left[\frac{-1}{2\pi r} v_\theta(r, \theta)e^{-in\theta}\right]_{-\pi}^{\pi} - \frac{in}{2\pi} \int_{-\pi}^{\pi} \frac{1}{r} v_\theta e^{-in\theta} d\theta$$

$$= 0 - \frac{in}{2\pi} \int_{-\pi}^{\pi} \frac{1}{r} v_\theta e^{-in\theta} d\theta$$

$$= \left[-\frac{in}{2\pi r} v(r, \theta)e^{-in\theta}\right]_{-\pi}^{\pi} + \frac{n^2}{2\pi} \int_{-\pi}^{\pi} \frac{1}{r} v(r, \theta)e^{-in\theta} d\theta$$

$$= \frac{n^2}{r} a_n.$$

Thus a_n satisfies the differential equation $(ra_n')' = (n^2/r)a_n$, which leads to $a_n = A_n r^n + B_n r^{-n}$ for some constants A_n and B_n $(n \neq 0)$. Since $v(r, \theta)$ is assumed to be bounded, we see that

$$|a_n(r)| = \left| \frac{1}{2\pi} \int_{-\pi}^{\pi} v(r, \theta)e^{-in\theta} d\theta \right| \leq 2 \max|v|$$

so that the coefficient B_n above must be 0, and $a_n(r) = A_n r^n$, for A_n a constant. Similarly $a_0(r) = A_0$. We can evaluate A_n by letting $r \to 1$, as follows.

$$A_n r^n = \frac{1}{2\pi} \int_{-\pi}^{\pi} v(r, \theta)e^{-in\theta} d\theta.$$

As $r \to 1$, $v(r, \theta) \to f(\theta)$ uniformly in θ, so that the integral converges to $(1/2\pi) \int_{-\pi}^{\pi} f(\theta)e^{-in\theta} d\theta$ as $r \to 1$, and we find

$$A_n = \frac{1}{2\pi} \int_{-\pi}^{\pi} f(\theta)e^{-in\theta} d\theta.$$

Thus for each fixed $r < 1$, the Fourier series of $v(r, \theta)$ is precisely the series for $u(r, \theta)$, the solution of the heat problem constructed through the Poisson kernel. Since this series converges uniformly to $u(r, \theta)$, the corollary of Theorem 1-2 asserts that $u(r, \theta) = v(r, \theta)$, which is what we were to show.

As an example of the application of this result, we can say that the temperature at the center of a disk with a steady state distribution is the average of the temperatures on the boundary, since

$$u(0, \theta) = A_0 = \frac{1}{2\pi} \int_{-\pi}^{\pi} f(\theta) \, d\theta$$

and $u(r, \theta)$ is the *only* solution of the heat problem.

Exercises. **1-15.** Prove that the steady state temperature in a disk is never higher than the highest temperature on the boundary. Prove that for a steady state temperature distribution in a disk, the coldest spot is on the boundary, unless the temperature is constant throughout the disk.

1-16. Let $\mathfrak{O}$ be an open connected set in the plane, and u a function defined in $\mathfrak{O}$ with second-order continuous derivatives, and with $u_{xx} + u_{yy} = 0$. (See Exercise 1-1.) Prove that if $|u|$ has a maximum in $\mathfrak{O}$, then u is constant. (This is the *maximum principle* for solutions of $u_{xx} + u_{yy} = 0$.)

1-17. Let (x_0, y_0) be a point in the plane, and suppose $u(x, y)$ is continuous for $(x - x_0)^2 + (y - y_0)^2 \leq R^2$, and $u_{xx} + u_{yy} = 0$ for $(x - x_0)^2 + (y - y_0)^2 < R^2$. Show that, for $0 \leq r < R$,

$$u(x_0 + r \cos \theta, y_0 + r \sin \theta) = \sum_{-\infty}^{\infty} a_n \left(\frac{r}{R}\right)^{|n|} e^{in\theta}$$

with

$$a_n = \frac{1}{2\pi} \int_{-\pi}^{\pi} u(x_0 + R \cos \theta, y_0 + R \sin \theta)e^{-in\theta} \, d\theta.$$

In particular,

$$u(x_0, y_0) = \frac{1}{2\pi} \int_{-\pi}^{\pi} u(x_0 + R \cos \theta, y_0 + R \sin \theta) \, d\theta.$$

This is a *mean value theorem* for solutions of Laplace's equation.

1-18. Rewrite the result of Exercise 1-17 to obtain the Poisson integral representation of a function u harmonic in a disk of radius R,

$$u(x_0 + r \cos \theta, y_0 + r \sin \theta) = \frac{1}{2\pi} \int_{-\pi}^{\pi} \frac{(1 - \rho^2)u(R \cos t, R \sin t) \, dt}{1 - 2\rho \cos (\theta - t) + \rho^2}$$
$$(\rho = r/R, 0 \leq r < R).$$

1-8. THE POINTWISE CONVERGENCE OF FOURIER SERIES

From a mathematical point of view, Poisson's theorem shows that any continuous function $f(\theta)$ of period 2π can be reconstructed from its Fourier series $\sum_{-\infty}^{\infty} a_n e^{in\theta}$, by inserting a "convergence factor" $r^{|n|}$ in the nth term, and taking the limit as $r \to 1-$. In this sense, at least, the series represents f. However, the simplest way for a series to represent a function f is for the series to converge to $f(\theta)$ for each value of θ. It is unfortunate that some continuous periodic functions are *not* represented by their Fourier series in this way. (An example is given in Section 8.2.8 of reference [17].) But if we assume a little more than mere continuity, for instance, differentiability, then the Fourier series of f does converge to f.

The first result of this nature was proved by Dirichlet, who showed in 1829 that if f is piecewise *monotone*, then the Fourier series of f actually converges to f at all points where f is continuous. We give in this section a slightly simpler version of Dirichlet's theorem.

The examination of pointwise convergence of Fourier series is similar to that of the series for $u(r, \theta)$; we convert the approximating sum into a single integral involving f. The symmetric partial sum S_N of the Fourier series of f is given by

$$S_N(\theta) = \sum_{-N}^{N} a_n e^{in\theta} = \frac{1}{2\pi} \int_{-\pi}^{\pi} f(t) \sum_{-N}^{N} e^{in(\theta-t)} \, dt,$$

where a_n is the nth Fourier coefficient of f. Thus the kernel relevant to pointwise convergence (corresponding to Poisson's kernel used previously) is

$$D_N(\theta) = \frac{1}{2\pi} \sum_{-N}^{N} e^{in\theta}$$

and we are interested in its behavior as $N \to \infty$. D_N is called *Dirichlet's kernel*.

Exercise. 1-19. Prove that

(i) $D_N(-\theta) = D_N(\theta)$.

(ii) $\int_{-\pi}^{\pi} D_N(\theta) \, d\theta = 1$.

(iii) $D_N(\theta) = \sin(N\theta + \theta/2)/2\pi \sin(\theta/2)$ (by induction).

Part (iii) of Exercise 1-19 shows that Dirichlet's kernel lacks two important features of Poisson's: it is not positive, and it does not converge to zero for $\theta \neq 0$; rather it oscillates more and more rapidly as $N \to \infty$. This oscillation is exploited by appeal to the following result, called Riemann's lemma.

Lemma. *If f is piecewise differentiable on the interval $a \leq \theta \leq b$, then $\lim_{A \to \infty} \int_a^b f(\theta) \sin(A\theta) \, d\theta = 0$.*

The assumption that f is piecewise differentiable is not essential, but it allows a trivial proof that covers the case we are interested in. Let $a_i < \theta < a_{i+1}$ $(i = 0, \ldots, n)$ be the intervals in which f has a continuous derivative. Then

$$\int_a^b f(\theta) \sin(A\theta) \, d\theta = \sum_{i=0}^n \int_{a_i}^{a_{i+1}}$$

$$= \frac{1}{A} \sum_0^n [f(a_i+) \cos(Aa_i) - f(a_{i+1}-) \cos(Aa_{i+1})]$$

$$+ \frac{1}{A} \sum_0^n \int_{a_i}^{a_{i+1}} f'(\theta) \cos(A\theta) \, d\theta$$

which $\to 0$ as $A \to \infty$.

This proof is very simple, but the reader should consider why the smoothness of f should make it easier to exploit the increasing rapidity of the oscillations of f; he should have in mind the graphs of $f(\theta) \sin A\theta$ for increasingly large values of A. Likewise, in the proof of Theorem 1-4 below, the reader should visualize the pictures corresponding to Figs. 1-3 and 1-4 illustrating Theorem 1-1. The argument is similar in many respects to that theorem.

Theorem 1-4. *Let f be periodic and piecewise differentiable. Then at each point θ the symmetric partial sums*

$$S_N(\theta) = \sum_{-N}^{N} a_n e^{in\theta}$$

converge to $\frac{1}{2}[f(\theta+) + f(\theta-)]$; *if f is continuous at θ, they converge to* $f(\theta)$.

Proof. By parts (i) and (ii) of Exercise 1-19, we have

$$f(\theta\pm) = 2 \int_0^\pi D_N(t) f(\theta\pm) \, dt$$

so

$$S_N(\theta) - \tfrac{1}{2}[f(\theta+) + f(\theta-)] = \int_0^\pi D_N(t)[f(\theta + t) - f(\theta+)] \, dt$$
$$+ \int_0^\pi D_N(t)[f(\theta - t) - f(\theta-)] \, dt.$$

These last two integrals are nearly identical, so we show only that

$$\int_0^\pi D_N(t)[f(\theta + t) - f(\theta+)] \, dt \to 0 \qquad \text{as} \quad N \to \infty. \quad (1\text{-}12)$$

Let δ_0 be any number less than π, and such that f has a continuous derivative f' for $\theta < t \le \theta + \delta_0$. Write, for $\delta \le \delta_0$,

$$\int_0^\pi [f(\theta + t) - f(\theta+)] D_N(t) \, dt = \int_0^\delta + \int_\delta^\pi$$
$$= I_1(\delta, N) + I_2(\delta, N).$$

Then if M is an upper bound for $|f'(t)|$ on $\theta < t \le \theta + \delta_0$, we have $|f(\theta + t) - f(\theta+)| \le Mt$ for $0 < t \le \delta_0$, and hence

$$|I_1(\delta, N)| \le \int_0^\delta Mt |D_N(t)| \, dt \le \frac{M}{2\pi} \int_0^\delta \frac{t}{\sin(t/2)} \, dt \le \frac{M\delta}{2}.$$

(See Exercise 1-20 below.) Thus for any given $\epsilon > 0$, $|I_1(\delta, N)| < \epsilon/2$ if $\delta \le \epsilon/M$, and this holds for all N. We conclude the proof of (1-12) by choosing $\delta = \epsilon/M$, say, and showing that $I_2(\delta, N) \to 0$ as $N \to \infty$; then for N sufficiently large we will have

$$|I_1(\delta, N) + I_2(\delta, N)| < \epsilon$$

proving (1-12). The convergence of I_2 is a direct consequence of the Riemann lemma: the function g of period 2π defined by

$$g(t) = 0 \qquad\qquad (-\pi < t \le \delta)$$

$$= \frac{f(\theta + t) - f(\theta+)}{2\pi \sin(t/2)} \qquad (\delta < t \le \pi)$$

is piecewise differentiable, and

$$I_2(\delta, N) = \int_{-\pi}^{\pi} g(t) \sin\left(Nt + \frac{t}{2}\right) dt.$$

This completes the proof of Theorem 1-4.

Exercises. **1-20.** Show $t \le \pi \sin(t/2)$ for $0 \le t \le \pi$.

1-21. Find the Fourier coefficients of the following functions, and write out their Fourier series in the form

$$a_0 + \sum_{1}^{\infty} (a_n e^{in\theta} + a_{-n} e^{-in\theta}).$$

Simplify the expressions by using

$$e^{in\theta} + e^{-in\theta} = 2 \cos(n\theta),$$

$$e^{in\theta} - e^{-in\theta} = 2i \sin(n\theta).$$

The functions all have period 2π.

    (a) $f(\theta) = \theta$ for $-\pi < \theta \le \pi$.
    (b) $g(\theta) = \theta^2$ for $-\pi \le \theta \le \pi$.
    (c) $h(\theta) = |\theta|$ for $-\pi \le \theta \le \pi$.
    (d) $k(\theta) = -1$ for $-\pi < \theta \le 0$, $k(\theta) = 1$ for $0 < \theta \le \pi$.

1-22. Apply Theorem 1-4 to evaluate various series of constants that are obtained from the examples of Exercise 1-21.

1-23. Suppose f is periodic and piecewise continuous. If $u(r, \theta)$ is given by (1-10), what is $\lim_{r\to 1} u(r, \theta)$? Give a proof.

1-24. If f is continuous and piecewise differentiable, and $g = f'$ at those points where f' exists, then the Fourier series of g is obtained from that of f by term-by-term differentiation. Prove this and show by examples that it fails when f is not continuous.

1-25. (**Summability**). A series $\Sigma_1^\infty u_n$, convergent or not, is called *Abel summable to L* if $\Sigma_1^\infty u_n r^n$ converges for $r < 1$ and $\lim_{r\to 1} \Sigma_1^\infty a_n r^n = L$. Thus Theorem 1-1 says that the Fourier series $a_0 + \Sigma_1^\infty (a_n e^{in\theta} + a_{-n} e^{-in\theta})$ of a continuous function f is Abel summable to $f(\theta)$. *Abel's theorem* states that if $\Sigma_1^\infty u_n$ converges to L, then it is Abel summable to L. A computationally simpler method of summation is Cesáro's: $\Sigma_1^\infty u_n$ is Cesáro summable to

L if the mean values $\sigma_n = (1/n)\Sigma_1^\infty s_j$ of the partial sums $s_n = \Sigma_1^n u_j$ converge to L, $\sigma_n \to L$.

(i) $\Sigma_1^\infty (-1)^n$ is Abel, and Cesáro, summable to $\frac{1}{2}$.

(ii)   If $\Sigma_1^\infty u_n = L$, then $\Sigma_1^\infty u_n$ is Cesáro summable to L.

(iii)   If a_j are the Fourier coefficients of f, $u_1 = a_0$, and $u_{n+1} = a_n e^{in\theta} + a_{-n} e^{-in\theta}$ for $n \geq 1$, then the Cesáro means are

$$\sigma_n(\theta) = \int_{-\pi}^{\pi} f(t) C_n(\theta - t)\, dt$$

$$nC_n = \sum_0^{n-1} D_k \qquad (D_k \text{ is the Dirichlet kernel}).$$

(iv)

$$\int_{-\pi}^{\pi} C_n(t)\, dt = 1, \qquad C_n(t) = \frac{\sin^2{(nt/2)}}{2n\pi \sin^2{(t/2)}}.$$

(The second formula comes from induction, with the formula $2 \sin(t/2) \sin(kt + t/2) = \cos kt - \cos(kt + t)$.)

(v)   (Fejér's theorem)   If f is continuous and has period 2π, then the Fourier series of f is Cesáro summable to f.

2

The Method of Separation
of Variables

This section discusses some trivial modifications of the Fourier series so far considered, with a view to various applications. The style will be very concise, which means that many statements are really simple exercises.

For a function f defined on an interval $(0, L)$, the *Fourier cosine series on* $(0, L)$ is

$$\tfrac{1}{2}A_0 + \sum_1^\infty A_n \cos \frac{nx\pi}{L}$$

with

$$(2\text{-}1)$$

$$A_n = \frac{2}{L} \int_0^L f(x) \cos \frac{nx\pi}{L}\, dx,$$

and the *Fourier sine series on* $(0, L)$ is

$$\sum_1^\infty B_n \sin \frac{n x \pi}{L} \quad \text{with} \quad B_n = \frac{2}{L} \int_0^L f(x) \sin \frac{n x \pi}{L} \, dx. \qquad (2\text{-}2)$$

The series (2-2) converges to zero if $x = 0$ or $x = L$. We shall find that if f is piecewise differentiable on $(0, L)$ (this supposes that $f(0+)$, $f(L-)$, $f'(0+)$, and $f'(L-)$ exist), then (2-1) and (2-2) converge to $\frac{1}{2}[f(x+) + f(x-)]$ for $0 < x < L$, and that (2-1) converges to $f(x)$ for $x = 0$ and $x = L$. Such properties of the sine and cosine series can generally be related to corresponding properties of the complete Fourier series, as below.

The series (2-1) and (2-2) differ from the standard Fourier series $\sum_{-\infty}^\infty a_n e^{in\theta}$ in two respects. First, the functions in the expansion have period $2L$, instead of 2π. Second, they use only sines, or only cosines, whereas the series of exponentials uses both. The first difference is easy to overcome, for if f has period $2L$, then the function g defined by

$$g(x) = f(xL/\pi)$$

has period 2π. The Fourier series $\sum_{-\infty}^\infty a_n e^{inx}$ for g yields a series $\sum_{-\infty}^\infty a_n e^{inx\pi/L}$ for f, called the *Fourier series of f on* $(-L, L)$. The coefficients are given by

$$a_n = \frac{1}{2L} \int_{-L}^L f(x) e^{-inx\pi/L} \, dx \qquad (2\text{-}3)$$

and the symmetric partial sums $\sum_{-N}^N a_n e^{inx\pi/L}$ converge to $\frac{1}{2}[f(x+) + f(x-)]$ if f is piecewise differentiable.

The second special feature of the series (2-1) and (2-2), their use of sines only, or of cosines only, is best understood by a consideration of *odd* and *even* functions. A function f is *even* if $f(x) = f(-x)$, and *odd* if $f(x) = -f(-x)$; this terminology is derived from the behavior of even and odd powers of x. The graph

$$\{(x, y) \colon y = f(x)\}$$

of a real-valued even function is symmetric about the line $y = 0$, and the graph of a real-valued odd function is symmetric about the origin. If g is any function, then

$$2g_{\text{even}}(x) = g(x) + g(-x)$$

defines an even function g_{even} and

$$2g_{\text{odd}}(x) = g(x) - g(-x)$$

defines an odd function g_{odd}. Since

$$g = g_{\text{even}} + g_{\text{odd}},$$

any function is the sum of an even and an odd function. When $g(x) = e^{ix}$, $g_{\text{even}}(x) = \cos x$ and $g_{\text{odd}}(x) = i \sin x$. Thus the series (2-1) represents an even function, and (2-2) an odd function. Conversely, if f is an even function of period $2L$, then the Fourier coefficients a_n of f satisfy $a_n = a_{-n}$, so that the symmetric partial sums of the Fourier series for f yield a series of the form (2-1):

$$\sum_{-N}^{N} a_n e^{in x \pi / L} = a_0 + \sum_{0}^{N} 2a_n \cos \frac{n x \pi}{L}.$$

If f is odd, its Fourier coefficients satisfy $a_n = -a_{-n}$, so the symmetric partial sums of the Fourier series yield a series of the form (2-2):

$$\sum_{-N}^{N} a_n e^{in x \pi / L} = \sum_{1}^{N} 2i a_n \sin \frac{n x \pi}{L}.$$

In the even case, we arrive at (2-1) by setting $A_n = 2a_n$ and using in (2-3) the fact that f is even; in the odd case, set $B_n = 2i a_n$ and use the fact that f is odd to convert (2-3) into the definition of B_n given in (2-2).

So far we have shown that the Fourier series of an even function of period $2L$ can be written in the form (2-1), and that of an odd function of period $2L$ in the form (2-2). Now suppose that f is defined only on the interval $0 < x \leq L$, and we want to obtain for it the expansion (2-1). Then it suffices to extend f to the interval $-L \leq x \leq L$ by making f even,

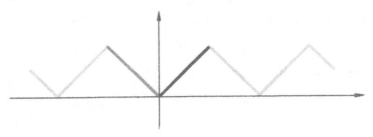

FIGURE 2-1. *Even periodic extension of $f(x) = x$ from interval*
$0 < x \leq L$.

that is, by setting

$$f(-x) = f(x) \qquad f(0) = f(0+)$$

and then over the whole line by making f periodic, that is, by
setting

$$f(x + 2L) = f(x).$$

A sketch of this even periodic extension is given in Fig. 2-1
for the case that $f(x) = x$ for $0 < x \leq L$. The properties
of the cosine series (2-1) for f follow by applying Theorem 1-4
to this extended function.

In the case of the sine series, the function f defined on
$0 < x \leq L$ is extended by making it *odd* and periodic, that is,
by setting

$$f(-x) = -f(x) \qquad f(0) = 0$$

$$f(x + 2L) = f(x)$$

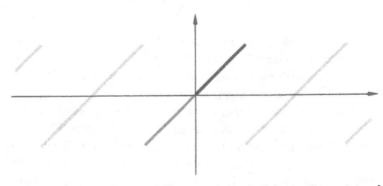

FIGURE 2-2. *Odd periodic extension of $f(x) = x$ from interval*
$0 < x < L$.

and the behavior of the sine series is obtained from this extension. (See Fig. 2-2.) In this way, all the claims made in the third paragraph of this section are verified.

2-2. THE VIBRATING STRING

The problem of heat in a disk is just one of many that can be solved by Fourier series. Section 2-3 will discuss some general features of this method of solution. Here we consider a particular problem that provides the physical motivation for some of the terminology of Fourier series, and some of the questions raised in the mathematical theory. Actually, the vibrating string problem considered here was one of the first to be analyzed with the help of sine series. An interesting account of some of the history of this question is given in Chapter 8 of Reference [17].

Consider a string which vibrates in the x-y plane, being stretched between the points $(0, 0)$ and $(L, 0)$. Let the displacement from the x axis at position x and time t be $u(x, t)$, that is, at time t the string passes through the point $(x, u(x, t))$. Then an argument something like the one in Section 1-1 leads to the equation

$$\frac{\rho \partial^2 u}{\partial t^2} = \frac{\tau \partial^2 u}{\partial x^2} \left[1 + \left(\frac{\partial u}{\partial x} \right)^2 \right]^{-1/2},$$

where ρ is the density of the string and τ the tension. This is at least reasonable, for it says that the force causing a section of the string to accelerate equals the tension times the curvature. This equation is a rather difficult one, and we attempt to get some insight into the situation by simplifying it. In particular, we suppose the string remains so nearly flat that the factor $[1 + (\partial u/\partial x)^2]^{1/2}$ can be replaced by the constant 1. Then the equation becomes

$$\frac{\tau \partial^2 u}{\partial x^2} = \frac{\rho \partial^2 u}{\partial t^2} \tag{2-4}$$

and it is the solutions of this equation that we shall study.

Since the ends $x = 0$ and $x = L$ are supposed fixed, we have

$$u(0, t) = 0 = u(L, t). \tag{2-5}$$

Finally, to describe the motion completely, we need to know the initial position and velocity of the string, that is,

$$u(x, 0) = f(x) \qquad (0 < x < L)$$
$$u_t(x, 0) = g(x) \qquad (0 < x < L), \tag{2-6}$$

where f and g are some given functions. (Actually, it took a number of people in the 18th century a number of years to decide that these were the relevant data: here we simply view this as a given, but reasonable, setting of the problem.)

Following the approach of Section 1-2 for the heat problem, we look first for nonzero product solutions $u(x, t) = X(x)T(t)$ satisfying (2-4) and (2-5), and try to superimpose them in such a way as to satisfy (2-6). Letting $u = XT$ in (2-4) we find $X''/X = \rho T''/\tau T = -c$, where c is some constant. The solutions for XT are then

$$(Ae^{ix\alpha} + Be^{-ix\alpha})(ae^{it\beta} + be^{-it\beta}) \tag{2-7}$$

with $\alpha = \sqrt{c}$, $\beta = \sqrt{c\tau/\rho}$. If $c < 0$, then α and β are imaginary, while if $c \geq 0$ we take these constants ≥ 0, for definiteness. Condition (2-5) makes

$$A + B = 0 \qquad \text{and} \qquad A(e^{iL\alpha} - e^{-iL\alpha}) = 0. \tag{2-8}$$

If α is imaginary, the last equation yields $A = 0$, so $u = 0$, and we discard this case. If α is real, then (2-8) yields $A \sin L\alpha = 0$; this can be satisfied with an $A \neq 0$ if $L\alpha$ is an integer multiple of π, $\alpha = n\pi/L \geq 0$. For these values of α, (2-7) yields the product solutions

$$u_n(x, t) = \left(\sin \frac{n\pi x}{L}\right)(a_n e^{itn\omega} + b_n e^{-itn\omega}) \tag{2-9}$$

with $\omega = (\pi/L)\sqrt{\tau/\rho}$.

Finally, we attempt to superimpose the solutions (2-9) in an infinite series

$$u(x, t) = \sum_1^\infty \left(\sin \frac{n\pi x}{L} \right) (a_n e^{itn\omega} + b_n e^{-itn\omega}) \quad (2\text{-}10)$$

satisfying (2-6). This leads, by formal term-by-term differentiation, to

$$\sum_1^\infty (a_n + b_n) \sin \frac{n\pi x}{L} = f(x)$$

$$\sum_1^\infty in\omega(a_n - b_n) \sin \frac{n\pi x}{L} = g(x). \quad (2\text{-}11)$$

Thus $a_n + b_n$ and $in\omega(a_n - b_n)$ are the coefficients in the Fourier sine series for f and g, respectively, and can be computed from Eq. (2-2). This completes the formal solution of the problem. Exercises 2-2, 2-3, and 2-4 below indicate how to recast the result in a form better suited to verifying conditions (2-4) and (2-6) rigorously.

The special solutions (2-9) have an interesting interpretation. Each one represents a vibration with frequency

$$\frac{n}{2L} \sqrt{\frac{\tau}{\rho}} \quad \text{cycles/sec.}$$

The lowest frequency

$$\frac{1}{2L} \sqrt{\frac{\tau}{\rho}}$$

is called the *fundamental*, and is usually the one we hear when the string is plucked. Apparently the fundamental increases with the tension, and decreases with the length and with the density. A glance in a piano confirms the last two effects (the bass strings are longer and heavier), and tuning any stringed instrument confirms the first. The fact that the other frequencies, called *harmonics*, are integral multiples of the fundamental is closely related to the fact that a stringed instrument sounds rather "pure," in contrast to the "jumbled" tone of a drum.   This terminology also explains why the subject of Fourier series is sometimes called harmonic analysis; the coefficients a_n and b_n in (2-10) determine the amplitudes of the various harmonics.

There is one further quantity associated with the vibrating string that plays a great role in analysis, namely the kinetic energy.   This is given, at any time t, by $\frac{1}{2} \int_0^L \rho u_t(x, t)^2 \, dx$. Differentiating, squaring, and integrating term by term yield for this integral

$$\left(-\frac{\rho L}{4} \right) \sum_1^\infty n^2 \omega^2 (a_n e^{itn\omega} - b_n e^{-itn\omega})^2.$$

Thus the total vibration is a sum of vibrations with frequencies $n\omega$, and the energy of the total vibration is the sum of the energies of the constituent vibrations.   This last fact seems a little surprising, since the constituents might be expected to interfere with each other.   We shall see later that these computations are justified even when the series for u_t does not converge uniformly in x.   The relevant result, called *Parseval's equality*, says that if a_n is the nth Fourier coefficient of a continuous function f of period $2L$, then

$$\int_{-L}^L |f|^2 = 2L \sum_{-\infty}^\infty |a_n|^2.$$

We establish this as part of Theorem 3-3.

Exercise.   **2-1.**   Solve the vibrating string problem with conditions (2-5) replaced by $u_x(0, t) = 0$ and $u_x(L, t) = 0$.   What physical situation does this describe?

The following exercises lead back to the original solution of the vibrating string problem offered by d'Alembert in 1747, and to the consequences of writing the solution in his form.

Exercises. **2-2.** Show that Eq. (2-10) can be rewritten in the form

$$u(x,\, t) = 2 \sum_{1}^{\infty} \sin \frac{n\pi x}{L} \left(A_n \cos n\omega t + B_n \sin n\omega t \right)$$

$$= \sum_{1}^{\infty} \left[A_n \sin n \left(\frac{\pi x}{L} + \omega t \right) - B_n \cos n \left(\frac{\pi x}{L} + \omega t \right) \right]$$

$$- \sum_{1}^{\infty} \left[A_n \sin n \left(\omega t - \frac{\pi x}{L} \right) - B_n \cos n \left(\omega t - \frac{\pi x}{L} \right) \right]$$

$$= v \left(\omega t + \frac{\pi x}{L} \right) - v \left(\omega t - \frac{\pi x}{L} \right)$$

assuming that these series converge. Here the A_n and B_n are constants related to the a_n and b_n of (2-10).

2-3. Show that if v is a function of one real variable, having period 2π and continuous first and second derivatives, then

$$u(x,\, t) = v \left(\omega t + \frac{\pi x}{L} \right) - v \left(\omega t - \frac{\pi x}{L} \right)$$

satisfies (2-4) and (2-5). Here $\omega = (\pi/L) \sqrt{\tau/\rho}$ as in (2-9).

2-4. Determine the function v in Exercise 2-3 from Eqs. (2-6). Show that the odd part of v is determined by f, and the even part by g. (Hint: The derivative of an even function is odd.) Show that v has continuous first and second derivatives if and only if the odd extension of period $2L$ of f has continuous first and second derivatives, and the same extension of g has continuous first derivatives.

2-5. (i) Determine the coefficients in Eq. (2-10) when $u_t(x,\, 0) = 0$ and $u(x,\, 0)$ has the graph in Fig. 2-3(a).

(ii) Same as part (i), but replace the graph of $u(x,\, 0)$ by that in Fig. 2-3(b).

(iii) For each of the answers in parts (i) and (ii) find, for the constituent vibrations with frequencies ω and 2ω, the amplitude and the time average of the kinetic energy.

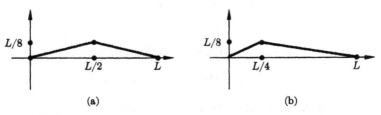

(a) (b)

FIGURE 2-3. *Initial positions of vibrating strings for Exercise 2-5.*

2-3. GENERALITIES ON THE METHOD
OF SEPARATION

The method applied in Section 1-2 to solve the problem of heat in a disk, and in Section 2-2 to solve that of the vibrating string, is called for more or less obvious reasons the *method of separation*. This approach can solve many (though certainly not all) particular problems involving partial differential equations. The present section discusses the method of separation in general, relying on a problem of heat transfer as an example.

Suppose a rod of length L lies on the x axis between $x = 0$ and $x = L$; and is insulated along its length and at both ends, so that heat is not lost, but may be conducted along the rod. Let the temperature at point x and time t be $u(x, t)$. Then if the specific heat of the rod per unit mass is c and the linear density is ρ, the heat between 0 and x is $\int_0^x c\rho u(s, t)\, ds$, and the rate of increase of this heat is $\int_0^x c\rho u_t(s, t)\, ds$. On the other hand, because of the insulation, heat can enter this section of the rod only at the right-hand end; if k is the conductivity of the rod, the rate at which heat enters there is $ku_x(x, t)$. Thus

$$\int_0^x c\rho u_t(s, t)\, ds = ku_x(x, t).$$

Differentiating with respect to x, we find

$$c\rho u_t(x, t) = ku_{xx}(x, t). \tag{2-12}$$

In order to determine a solution, recall that the ends were to be insulated, so that

$$u_x(0, t) = 0 \qquad u_x(L, t) = 0 \qquad \text{and}$$
$$u(x, t) \quad \text{is bounded for} \quad 0 \le t < \infty. \tag{2-13}$$

We specify the initial temperature distribution

$$u(x, 0) = f(x) \qquad (0 < x < L). \tag{2-14}$$

In the method of separation, we try to obtain a solution as a sum of products, $\Sigma A_n X_n(x) T_n(t)$, where A_n is constant, $X_n(x)T_n(t)$ satisfies some of the conditions (i)–(iv), and the series satisfies all the conditions. These considerations determine which conditions can be imposed on the products: those which, if satisfied by all the X_nT_n, are also satisfied by $\Sigma\, A_nX_nT_n$. To put it more simply, they are those which, if satisfied by two functions v and w, are also satisfied by $Av + Bw$ Such a condition we call linear homogeneous: *the condition $C(u)$ on functions u is called* linear homogeneous *if and only if the truth of $C(v)$ and $C(w)$ implies the truth of $C(Av + Bw)$ for all constants A and B.* Thus the condition $C(u)$ given by the first equation (2-13) above is linear homogeneous; for if $v_x(0,\ t) = 0$ and $w_x(0,\ t) = 0$ (that is, if $C(v)$ and $C(w)$ are true) then $(Av + Bw)_x(0,\ t) = Av_x(0,\ t) + Bw_x(0,\ t) = 0$ (that is, $C(Av + Bw)$ is true). Condition (2-12) is likewise linear homogeneous. Condition (2-14), on the other hand, is not linear homogeneous when $f \neq 0$; for if $v(x, 0) = f(x)$ and $w(x, 0) = f(x)$, then $Av(x, 0) + Bw(x, 0) = Af(x) + Bf(x)$, while condition (2-14) applied to the function $Av + Bw$ requires $Av(x, 0) + Bw(x, 0) = f(x)$.

Thus for the problem at hand, we look for product solutions $X(x)T(t)$ satisfying the linear homogeneous conditions (2-12) and (2-13), and try to superimpose these in such a way as to satisfy (2-14). This is the basis of the method of separation.

Exercise. 2-6.
    (a)   Show that conditions (2-12) and (2-13) are linear homogeneous.
    (b)   Find all products XT satisfying these linear homogeneous conditions.
    (c)   What happens to each product solution as $t \to +\infty$?
    (d)   Complete the solution in case $u(x, 0) = x$.
    (e)   What is $\lim_{t\to +\infty} u(x,\ t)$? Does this make sense physically?
    (f)   How much heat is in the rod at time t? Is this reasonable?

Even when it is not possible to solve the problem in the way suggested above, further manipulations may yield the solution. Suppose, for instance, that the rod in the above problem is carrying current, so that Eq. (2-12) is replaced by

$$c\rho u_t(x,\ t) = ku_{xx}(x,\ t) + H, \tag{2-15}$$

where H is the amount of heat per second per unit length generated by the current in the rod. Instead of considering a rod insulated at the ends, we suppose that the ends are maintained at a reference temperature T_0, so that (2-13) is replaced by

$$u(0, t) = T_0 \qquad u(L, t) = T_0. \qquad (2\text{-}16)$$

To complete the specifications, suppose that at $t = 0$

$$u(x, 0) = T_0. \qquad (2\text{-}17)$$

Here we find none of the conditions is linear homogeneous (unless $T_0 = 0$), so that the above outline is no help. Even if the temperature scale were adjusted so that $T_0 = 0$, the inhomogeneity of (2-15) would spoil the method.

However, the problem can be solved by a preliminary reduction suggested by the physical situation. In this particular problem we expect the temperature to approach a "steady state" $s(x)$, $u(x, t) \rightarrow s(x)$ as $t \rightarrow +\infty$. The conditions on $v(x, t) = u(x, t) - s(x)$ are more likely to be homogeneous than those on $u(x, t)$. The function $v(x, t)$ is thought of as the "transient" temperature distribution.

To find the probable steady state $s(x)$, note that for $s(x)$ all time derivatives would be zero, so $ks''(x) = -H$, and $s(0) = s(L) = T_0$. The solution of this differential equation is $s(x) = T_0 + Hx(L - x)/2k$. Setting now $u(x, t) = v(x, t) + s(x)$, we find for v the conditions

$$c\rho v_t = kv_{xx}$$

$$v(0, t) = 0$$

$$v(L, t) = 0 \qquad (2\text{-}18)$$

$$v(x, 0) = T_0 - s(x) = Hx(x - L)/2k.$$

Exercises. **2-7.** Find v satisfying Eqs. (2-18).

2-8. Estimate, in terms of c, k, and L, how long it takes before $v(x, t) < L^2H/10k$ for $0 < x < L$; that is, when the transient becomes relatively small compared with the variation in the steady state between $x = 0$ and $x = L/2$.

The above discussion of the separation method is far from complete. For instance, there are many other tricks, such as the above reduction to equations for the transient, to extend the applicability of the method (see Exercise 2-11, for example); and separation often leads to expansions other than a Fourier series, for instance, to Fourier integrals, Hermite and Legendre series, and series or integrals of Bessel functions, to name a few. More methods and examples along these lines are to be found in books that treat partial differential equations of physics of engineering, or specifically boundary value problems.

The reader should not think that most problems can be solved by this method, for its success depends on having very special regions and equations. However, many of these special cases are the most interesting ones, and the rather explicit answers available for them provide some insight into the solution of more general problems.

Exercises. **2-9.** The steady state temperature $u(x, y)$ in a rectangular plate $0 \leq x \leq L$, $0 \leq y \leq M$, is sought, under the condition that the edge $x = 0$ is maintained at zero degrees, $x = L$ is kept at $u(L, y) = y$ degrees, and the edges $y = 0$ and $y = M$ are insulated. The appropriate differential equation is $u_{xx} + u_{yy} = 0$. (See Exercise 1-1).

(a) Express the boundary conditions mathematically, and identify the linear homogeneous ones.

(b) Find all products satisfying the differential equation and the linear homogeneous boundary conditions.

(c) Find a series for $u(x, y)$.

2-10. Suppose the plate of the previous exercise is not in a steady state, so the temperature is $u(x, y, t)$. Suppose the same boundary conditions prevail as in Exercise 2-9, and that at $t = 0$ the temperature is $u(x, y, 0) = f(x, y)$.

(a) Show that the differential equation for u is $c\rho u_t = k(u_{xx} + u_{yy})$, where c is specific heat, ρ is mass per unit area, and k is conductivity.

(b) Find all products XYT satisfying the appropriate linear homogeneous conditions. For simplicity, let $c = \rho = k = 1$, and $L = M = \pi$.

(c) What happens to these products as $t \to +\infty$? (Only those which are bounded as $t \to +\infty$ are used in the actual solution.)

(d) Find a series for $u(x, y, t)$ in case $u(x, y, 0) = y$.

2-11. Suppose that in Exercise 2-9 the condition $u(0, y) = 0$ is replaced by $u(0, y) = -y$, the others remaining the same. The solutions of this can be obtained as the sum of the solution of Exercise 2-9 and a similar solution v satisfying $v(0, y) = -y$, $v(L, y) = 0$.

3

Some Applications of
Poisson's Theorem

One of the classical results of mathematical analysis is the
Weierstrass approximation theorem: a continuous function
on a closed finite interval can be uniformly approximated on
that interval by a polynomial.   This and similar approximation theorems have since become standard tools of mathematical analysis.   Here we deduce Weierstrass' theorem, and
some others, and as an application establish Parseval's equality, mentioned briefly at the end of Section 2-2.

3-1. *UNIFORM APPROXIMATION*

Any finite sum of the form $\Sigma_{-N}^{N} b_n e^{in\theta}$ is called a trigonometric
polynomial of degree $\leq N$, since $e^{\pm in\theta} = (\cos \theta \pm i \sin \theta)^n$.
Clearly, every trigonometric polynomial is a continuous func-
tion of period 2π, and it is equally clear that there are continuous periodic functions that are not such polynomials, for

example, the periodic extension of f: $f(x) = |x|$ ($|x| \leq \pi$). However, no continuous periodic function is far from the trigonometric polynomials, in the following sense.

Theorem 3-1. *Let f be a continuous function of period 2π. Then for each $\epsilon > 0$ there is a trigonometric polynomial P such that $|P(\theta) - f(\theta)| < \epsilon$ for all θ.*

Proof. By Poisson's theorem (Theorem 1-1), there is an $r < 1$ such that $|u(r, \theta) - f(\theta)| < \epsilon/2$, where

$$u(r, \theta) = \sum_{-\infty}^{\infty} a_n r^{|n|} e^{in\theta}$$

and

$$a_n = \frac{1}{2\pi} \int_{-\pi}^{\pi} f(\theta) e^{-in\theta} \, d\theta.$$

Since $|a_n| \leq \max|f|$, we have

$$\left| \sum_{|n| \geq N} a_n r^{|n|} e^{in\theta} \right| \leq 2 \max|f| \frac{r^N}{1 - r}. \tag{3-1}$$

If we choose N so that the right-hand side of (3-1) is $< \epsilon/2$, we have

$$\left| f(\theta) - \sum_{|n| \leq N} a_n r^{|n|} e^{in\theta} \right| < |f(\theta) - u(r, \theta)| + \frac{\epsilon}{2} < \epsilon$$

and Theorem 3-1 is proved.

Since the functions $e^{in\theta}$ can be uniformly approximated by their Taylor series, we can obtain Weierstrass' theorem from Theorem 3-1.

Theorem 3-2. *Let f be a continuous function on the finite closed interval $[a, b]$. Then given $\epsilon > 0$ there is a polynomial P such that $|P - f| < \epsilon$ on $[a, b]$.*

Proof. Let c be greater than $\max(|a|, |b|)$. Then f can be extended to a continuous function on $[-c, c]$ by adjoining to the graph of f the line segments from the point $(-c, 0)$ to the

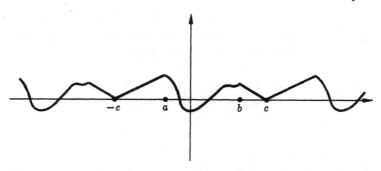

FIGURE 3-1. *Continuous periodic extension of f from $a \leq x \leq b$.*

point $(a, f(a))$, and from $(b, f(b))$ to $(c, 0)$.   Then f can be further extended so as to be continuous and have period $2c$; call this extension F.   (See Fig. 3-1.)   Let $g(\theta) = F(c\theta/\pi)$, so g has period 2π.   Then choose a trigonometric polynomial

$$Q(\theta) = \sum_{-N}^{N} a_n e^{in\theta} \quad \text{with} \quad |Q - g| < \frac{\epsilon}{2}.$$

Finally, approximate $e^{in\theta}$ by a polynomial $P_n(\theta)$ (say, a finite number of terms of its Taylor series) so that

$$|P_n(\theta) - e^{in\theta}| < \frac{\epsilon}{|a_n|(4N + 2)} \quad \text{for} \quad |\theta| \leq c.$$

Then

$$P(\theta) = \sum_{-N}^{N} a_n P_n(\theta)$$

satisfies

$$|g(\theta) - P(\theta)| \leq |g(\theta) - Q(\theta)| + |Q(\theta) - P(\theta)| < \frac{\epsilon}{2} + \frac{\epsilon}{2} = \epsilon$$

so

$$\left| f(x) - P\left(\frac{\pi x}{c}\right) \right| < \epsilon \quad \text{for} \quad a \leq x \leq b.$$

Theorems 3-2 and 3-1 are sometimes called *Weierstrass' first and second approximation theorems*, respectively.

Exercises.   **3-1.**   Let f be a continuous function on $[-1, 1]$.   If f is even, f can be approximated uniformly by an even polynomial. Prove a corresponding result if f is odd.

3-2. Let f be continuous on the finite closed interval $[a, b]$. What can be said if $\int_a^b x^n f(x)\, dx = 0$ for every integer $n \geq 0$? Prove your result.

3-3. Let f be continuous on $[-1, 1]$. What can be said if $\int_{-1}^1 x^{2n} f(x)\, dx = 0$ for every integer $n \geq 0$? What if $\int_{-1}^1 x^n f(x)\, dx = 0$ for every integer $n > 0$?

3-2. *LEAST SQUARES APPROXIMATION*

The approximation we consider here is not so easy to visualize graphically as the uniform approximation of the previous section, but it is nonetheless important, largely because it is well adapted to calculations. A simple case of such an approximation is the least squares method of fitting a straight line to a given set of points.

Theorem 3-3. *Let a_n be the Fourier coefficients of the continuous function f of period 2π, and*

$$S_N(\theta) = \sum_{-N}^{N} a_n e^{in\theta}.$$

Then

$$\int_{-\pi}^{\pi} |S_N - f|^2 \to 0 \qquad \text{as} \quad N \to \infty. \tag{3-2}$$

Further, if

$$t_N(\theta) = \sum_{-N}^{N} b_n e^{in\theta}$$

with arbitrary coefficients b_n, then

$$\int_{-\pi}^{\pi} |t_N - f|^2 > \int_{-\pi}^{\pi} |S_N - f|^2 \tag{3-3}$$

unless $b_n = a_n$ for $-N \leq n \leq N$.

Finally, Parseval's equality holds:

$$2\pi \sum_{-\infty}^{\infty} |a_n|^2 = \int_{-\pi}^{\pi} |f|^2. \tag{3-4}$$

Before discussing the proof, notice how much better this sort of approximation fits Fourier series than pointwise or uniform approximation. We have not been able to show that $S_N \to f$ uniformly, or even that $S_N(\theta) \to f(\theta)$ for each θ; in fact neither of these statements is true for an arbitrary continuous f. By contrast, (3-2) holds when f is continuous; it even holds under the weaker assumptions that $\int_{-\pi}^{\pi} |f|^2$ exists in some sense, for instance, in the sense of improper Riemann integrals, but we shall not prove this. (See Exercises 3-11 and 3-12, however.) Moreover, the partial sums S_N of the Fourier series are the best possible approximation by trigonometric polynomials of degree $\leq N$, as (3-3) shows. Finally, the term-by-term integration implied in (3-4) is justified, even though the series $\Sigma\, a_n e^{in\theta}$ may not converge uniformly.

For a physical interpretation, recall from Section 2-2 that if $f(x, t)$ denotes the velocity at time t of the portion near the point x of a vibrating string of density ρ stretched between $x = -\pi$ and $x = \pi$, then the kinetic energy of the string is

$$\frac{\rho}{2} \int_{-\pi}^{\pi} f(x, t)^2 \, dx.$$

In this situation, the inequality (3-3) says the following: If the vibration described by f is to be approximated by a combination of trigonometric vibrations of the form $\Sigma_{-N}^{N}\, b_n(t) e^{ixn}$, then the approximation that accounts for the energy in the most efficient way is the one that sets the $b_n(t)$ equal to the Fourier coefficients.

We begin the proof with the inequality (3-3). Here

$$\int_{-\pi}^{\pi} |t_N - f|^2$$

$$= \int_{-\pi}^{\pi} \left[\sum_{-N}^{N} b_n e^{in\theta} - f(\theta) \right] \left[\sum_{-N}^{N} \bar{b}_m e^{-im\theta} - \bar{f}(\theta) \right] d\theta$$

$$= 2\pi \sum |b_n|^2 - 2\pi \sum b_n \bar{a}_n - 2\pi \sum \bar{b}_m a_m + \int |f|^2,$$

and substituting a_n for b_n yields

$$\int_{-\pi}^{\pi} |S_N - f|^2 = \int |f|^2 - 2\pi \sum_{-N}^{N} |a_n|^2. \qquad (3\text{-}5)$$

Thus

$$\int |t_N - f|^2 - \int |S_N - f|^2 = 2\pi (\Sigma |b_n|^2 - \Sigma b_n \bar{a}_n - \Sigma \bar{b}_m a_m \\ + \Sigma |a_n|^2)$$

$$= 2\pi \Sigma (b_n - a_n)(\bar{b}_n - \bar{a}_n) > 0$$

unless $b_n = a_n$ for $-N \leq n \leq N$. This proves (3-3); notice that the proof uses nothing more than the algebra of complex numbers, the fact that $\int_{-\pi}^{\pi} e^{i(n-m)\theta} \, d\theta = 0$ when $n \neq m$, and the definition of the Fourier coefficients a_n.

To prove the relation (3-2), suppose $\epsilon > 0$ is given, and choose a trigonometric polynomial

$$P(\theta) = \sum_{-N}^{N} b_n e^{in\theta} \quad \text{with} \quad |P - f|^2 < \frac{\epsilon}{2\pi};$$

this is possible by Theorem 3-1. Then

$$\int |S_N - f|^2 \leq \int |P - f|^2 < \epsilon$$

by (3-3), so again by (3-3)

$$\int |S_M - f|^2 \leq \int |S_N - f|^2 < \epsilon$$

if $M \geq N$.

Finally, for Parseval's equality we have from Eq. (3-5) that

$$2\pi \sum_{-N}^{N} |a_n|^2 \leq \int_{-\pi}^{\pi} |f|^2$$

which is known as *Bessel's inequality*. From this we have the convergence of $\Sigma_{-N}^{N} |a_n|^2$, and in fact

$$2\pi \sum_{-\infty}^{\infty} |a_n|^2 \leq \int_{-\pi}^{\pi} |f|^2$$

which is half of Parseval's equality. For the other half, take a trigonometric polynomial P of degree M with

$$\int_{-\pi}^{\pi} |P - f|^2 < \epsilon.$$

Then by (3-3) and (3-5) again,

$$2\pi \sum_{-M}^{M} |a_n|^2 = \int_{-\pi}^{\pi} |f|^2 - \int_{-\pi}^{\pi} |S_M - f|^2 \geq \int_{-\pi}^{\pi} |f|^2 - \epsilon$$

so that for $N \geq M$

$$\int_{-\pi}^{\pi} |f|^2 - \epsilon \leq 2\pi \sum_{-N}^{N} |a_n|^2 \leq \int_{-\pi}^{\pi} |f|^2.$$

Exercises. **3-4.** Suppose f is piecewise continuous. Prove Bessel's inequality. Show that the Fourier coefficients a_n of f converge to zero as $n \to \infty$. (Compare this with the Riemann lemma.)

3-5. Evaluate $\sum_1^\infty n^{-4}$ from the Fourier series of the periodic extension of $f : f(\theta) = \theta^2$ for $-\pi < \theta \leq \pi$.

3-6. Let f be a continuous positive function of period 2π. Express the area inside the curve $r = f(\theta)$ in terms of the Fourier coefficients of f. (Here r and θ are polar coordinates.)

3-7. Let $f(\theta) = \theta^2$ for $-\pi \leq \theta \leq \pi$. Choose the coefficient a_0 to minimize each of the following expressions:

    (i) $|f(0) - a_0|$.

    (ii) $\int_{-\pi}^{\pi} |f(\theta) - a_0|\, d\theta$.

    (iii) $\int_{-\pi}^{\pi} |f(\theta) - a_0|^2\, d\theta$.

    (iv) $\max_{-\pi \leq \theta \leq \pi} |f(\theta) - a_0|$.

Note that the four choices of a_0 are all different, while each gives the best approximation of f by a constant, in some sense.

3-3. INNER PRODUCTS AND SCHWARZ'S INEQUALITY

This section investigates the "mean square" integral further, particularly in analogy to the familiar algebra of vectors in the plane.

Recall that if $\mathbf{u} = u_1\mathbf{i} + u_2\mathbf{j}$ and $\mathbf{v} = v_1\mathbf{i} + v_2\mathbf{j}$ are vectors in the plane, then their inner product $(\mathbf{u}, \mathbf{v})$ is defined as $(\mathbf{u}, \mathbf{v}) = u_1 v_1 + u_2 v_2$, and $|\mathbf{u}| = (\mathbf{u}, \mathbf{u})^{1/2} = (u_1{}^2 + u_2{}^2)^{1/2}$ is the length of $\mathbf{u}$. (This inner product is often denoted $\mathbf{u} \cdot \mathbf{v}$ and called the *dot product*.) The geometric interpretation of the inner product is given by $(\mathbf{u}, \mathbf{v}) = |\mathbf{u}|\, |\mathbf{v}| \cos \phi$, where ϕ is the angle between $\mathbf{u}$ and $\mathbf{v}$. Since $\cos \phi = 0$ when $\phi = \pi/2$, $\mathbf{u}$ and

v are called *orthogonal* when $(\mathbf{u}, \mathbf{v}) = 0$. Corresponding remarks apply to n-dimensional real vectors $\mathbf{u} = \Sigma_1^n u_j \mathbf{i}_j$, with

$$(\mathbf{u}, \mathbf{v}) = \Sigma_1^n u_j v_j = |\mathbf{u}| \, |\mathbf{v}| \cos \phi, \quad |\mathbf{u}|^2 = (\mathbf{u}, \mathbf{u}).$$

Thus, since $|\cos \phi| \leq 1$, we have $|(\mathbf{u}, \mathbf{v})| \leq |\mathbf{u}| \, |\mathbf{v}|$, which is called *Schwarz's inequality;* and the parallelogram law for addition of vectors leads us to call $|\mathbf{u} + \mathbf{v}| \leq |\mathbf{u}| + |\mathbf{v}|$ the *triangle inequality.*

The concepts of inner product and length are basic in analytic geometry, at least partly because they are easy to compute and have simple geometric interpretations. It is thus fortunate that these same concepts are available for large classes of functions, and here too they are basic tools of analysis. In particular, for piecewise continuous, real functions f and g on a finite interval $[a, b]$, we define the inner product

$$(f, g) = \frac{1}{b - a} \int_a^b fg \tag{3-6}$$

and the length, or norm,

$$\|f\| = (f, f)^{1/2}. \tag{3-7}$$

(The notation $\|f\|$ is chosen to avoid confusion with the function $|f|$, the absolute value of f, defined by $|f|(x) = |f(x)|$.) Theorem 3-4 shows that this inner product behaves very much like the familiar "dot product" in the plane.

Theorem 3-4. *Let f and g be real, piecewise continuous functions. If c_1 and c_2 are real constants, then*

$$(c_1 f_1 + c_2 f_2, g) = c_1(f_1, g) + c_2(f_2, g). \tag{3-8}$$

Further,

$$(f, g) = (g, f) \tag{3-9}$$

and

$$(f, f) \geq 0 \tag{3-10}$$

with equality only when $f = 0$ at all but finitely many points; and

$$|(f, g)| \leq \|f\| \cdot \|g\| \tag{3-11}$$

$$\|f + g\| \leq \|f\| + \|g\|. \tag{3-12}$$

Proof. The statements (3-8) and (3-9) are simple computations. Schwarz's inequality (3-11) can be reduced to the corresponding inequality in n-dimensional space, by approximating the integrals with Riemann sums. However, there is a simpler and more interesting proof. Let λ denote a real variable, and consider the fact that (3-8)–(3-10) imply.

$$0 \leq (f + \lambda g, f + \lambda g) = \|f\|^2 + 2\lambda(f, g) + \lambda^2 \|g\|^2.$$

Thus the graph of this quadratic lies entirely above the λ axis, and there is at most one real root of the equation

$$\|g\|^2 \lambda^2 + 2(f, g)\lambda + \|f\|^2 = 0.$$

From this and the formula for the roots of a quadratic equation, it follows that the discriminant $4(f, g)^2 - 4\|f\|^2 \|g\|^2$ is less than or equal to zero; this is equivalent to the inequality (3-11). Finally, (3-12) follows from the previous parts:

$$\begin{aligned}(f + g, f + g) &= \|f\|^2 + 2(f, g) + \|g\|^2 \\ &\leq \|f\|^2 + 2\|f\| \cdot \|g\| + \|g\|^2 \\ &= (\|f\| + \|g\|)^2.\end{aligned}$$

Exercise. 3-8. Prove Schwarz's and the triangle inequalities in n space, that is, $(\Sigma_1^n u_j v_j)^2 \leq (\Sigma_1^n u_j^2)(\Sigma_1^n v_j^2)$ and $[\Sigma_1^n (u_j + v_j)^2]^{1/2} \leq (\Sigma u_j^2)^{1/2} + (\Sigma v_j^2)^{1/2}$.

Since Theorem 3-4 applies only to real functions, it is not quite adequate for our theory, where the complex functions $e^{in\theta}$ are prominent. For complex functions we make a new definition of the inner product that agrees with (3-6) when the functions are real, namely

$$(f, g) = \frac{1}{b - a} \int_a^b f\bar{g}. \tag{3-13}$$

The complex conjugate in the integral has the effect that $(f, f) = \int |f|^2 \geq 0$, so we can again set $\|f\| = (f, f)^{1/2}$. This inner product behaves almost like the real form (3-6).

Theorem 3-5. *If f and g denote complex functions and c a complex number, then*

$$(c_1 f_1 + c_2 f_2, g) = c_1(f_1, g) + c_2(f_2, g) \qquad (3\text{-}14)$$

$$(f, g) = \overline{(g, f)} \qquad (3\text{-}15)$$

$$(f, f) \geq 0 \qquad (3\text{-}16)$$

$$|(f, g)| \leq \|f\| \cdot \|g\| \qquad (3\text{-}17)$$

and

$$\|f + g\| \leq \|f\| + \|g\|.$$

The proof follows that of Theorem 3-4, except for (3-17). Proceeding as before, we find at most one real root of

$$\|g\|^2 \lambda^2 + [(f, g) + (g, f)]\lambda + \|f\|^2$$
$$= \|g\|^2 \lambda^2 + 2 \operatorname{Re}(f, g)\lambda + \|f\|^2 = 0$$

so $|\operatorname{Re}(f, g)| \leq \|f\| \cdot \|g\|$. However, we can pick a real number α such that $e^{i\alpha}(f, g) = |(f, g)|$, and then have

$$|(f, g)| = \operatorname{Re}(e^{i\alpha}f, g) \leq \|e^{i\alpha}f\| \cdot \|g\| = \|f\| \cdot \|g\|.$$

Notice that (3-14) and (3-15) imply that $(f, cg) = \bar{c}(f, g)$, *not* $c(f, g)$.

On the basis of Theorem 3-5, we can (and do) apply many of the concepts of 3-space to the piecewise continuous functions on a finite interval. (The restriction to piecewise continuous functions is for the sake of simplicity; the same remarks apply to any class of functions for which the integrals $\int f\bar{g}$ make sense.) In particular, when $a = -\pi$ and $b = \pi$, the functions $e_n(\theta) = e^{in\theta}$ are mutually orthogonal (that is, $(e_n, e_m) = 0$ if $n \neq m$) and have unit length (that is, $\|e_n\|^2 = (1/2\pi) \int_{-\pi}^{\pi} e^{in\theta} e^{-in\theta}\, d\theta = 1$). In this they resemble the unit vectors $\mathbf{i}, \mathbf{j}, \mathbf{k}$ in 3-space, and indeed the resemblance goes much further. For instance, the components u_1, u_2, u_3 of the vector $\mathbf{u} = u_1\mathbf{i} + u_2\mathbf{j} + u_3\mathbf{k}$ are given by $u_1 = (\mathbf{u}, \mathbf{i})$, $u_2 = (\mathbf{u}, \mathbf{j})$, and $u_3 = (\mathbf{u}, \mathbf{k})$, while the coefficients a_n of the Fourier series

$$\sum_{-\infty}^{\infty} a_n e^{in\theta}$$

for f are given by $a_n = (f, e_n)$.   (See Eq. (3-13).)   The three-dimensional "Pythagorean theorem"

$$\|\mathbf{u}\|^2 = \sum_1^3 u_j^2$$

corresponds to the Parseval equality

$$\|f\|^2 = \frac{1}{2\pi} \int_{-\pi}^{\pi} |f|^2 = \sum_{-\infty}^{\infty} |a_n|^2.$$

One important difference, however, is the following.   Given any $\mathbf{u}$ in 3-space, we have the triple (u_1, u_2, u_3) of components of $\mathbf{u}$, and conversely, given any triple (u_1, u_2, u_3) of real numbers there is a vector $\mathbf{u} = u_1\mathbf{i} + u_2\mathbf{j} + u_3\mathbf{k}$ with these components.   With the Fourier series we again find, for each piecewise continuous function f, Fourier coefficients a_n satisfying

$$\sum_{-\infty}^{\infty} |a_n|^2 < \infty. \tag{3-18}$$

But the converse fails, for there are sequences satisfying (3-18) that are not the Fourier coefficients of any piecewise continuous function.   The difficulty is that our class of functions is not quite large enough; this can be rectified, but not in a page or two.   One of the most significant advances in modern analysis was the introduction by H. Lebesgue of a more powerful type of integration, applying to a class of functions larger than those considered up to that time.   This innovation made possible, among other things, the *Riesz-Fisher theorem:* For every sequence $\{a_n\}_{n=-\infty}^{\infty}$ satisfying (3-18), there is a function f such that $|f|^2$ and $f\bar{e}_n$ are integrable in the sense of Lebesgue, and such that

$$\frac{1}{2\pi} \int_{-\pi}^{\pi} f\bar{e}_n = a_n \quad \text{and} \quad \frac{1}{2\pi} \int_{-\pi}^{\pi} |f|^2 = \sum_{-\infty}^{\infty} |a_n|^2.$$

The Lebesgue integral has many further advantages, and consequently all advanced work in analysis, including half the books in our bibliography, is based on it. It is fortunate that many aspects of Fourier series such as those treated in this little book do not depend in an essential way on the Lebesgue theory, but no account can be reasonably complete without it. However, we may console the reader by reminding him that our subject is appealing not because it is complete in itself, but rather because it suggests the background, and sometimes the solution, of some important problems and theories developed in the past hundred years or so.

Exercises. **3-9.** Prove Schwarz's and the triangle inequalities for infinite sequences: If

$$\sum_{-\infty}^{\infty} |a_n|^2 < \infty \quad \text{and} \quad \sum_{-\infty}^{\infty} |b_n|^2 < \infty$$

then

(i) $\Sigma_{-\infty}^{\infty} a_n b_n$ converges absolutely and

$$\left| \sum_{-\infty}^{\infty} a_n b_n \right|^2 \leq \left(\sum_{-\infty}^{\infty} |a_n|^2 \right) \left(\sum_{-\infty}^{\infty} |b_n|^2 \right)$$

and

(ii) $\left(\sum_{-\infty}^{\infty} |a_n + b_n|^2 \right)^{\frac{1}{2}} \leq \left(\sum_{-\infty}^{\infty} |a_n|^2 \right)^{\frac{1}{2}} + \left(\sum_{-\infty}^{\infty} |b_n|^2 \right)^{\frac{1}{2}}.$

3-10. Denote by l^2 the space of sequences $a = \{a_n\}$ such that $\Sigma_{-\infty}^{\infty} |a_n|^2 < \infty$. Then by the previous exercise $(a, b) = \Sigma_{-\infty}^{\infty} a_n \bar{b}_n$ is defined, and if $\|a\|^2 = \Sigma |a_n|^2$, then $\|a + b\| \leq \|a\| + \|b\|$.

A Cauchy sequence in l^2 is a sequence $\{a^m\}_{m=1}^{\infty}$ of members of l^2 such that $\|a^m - a^p\| \to 0$ as $m, p \to \infty$. (Note that each a^m is a sequence, $a^m = \{a_n^m\}_{n=-\infty}^{\infty}$.) Prove that if $\{a^m\}$ is a Cauchy sequence in l^2, then there is a member a of l^2 such that $\|a^m - a\| \to 0$; that is, every Cauchy sequence converges to a limit in l^2.

3-11. Suppose f is piecewise continuous. Show that for each $\epsilon > 0$ there is a continuous function g of period 2π such that $\int_{-\pi}^{\pi} |f - g|^2 < \epsilon$. Show that there is a trigonometric polynomial P such that $\int_{-\pi}^{\pi} |f - P|^2 < \epsilon$. Finally, show that Theorem 3-3 holds when f is piecewise continuous.

3-12. Show that Theorem 3-3 holds when f is only assumed Riemann integrable.

3-13. Suppose f and g are continuous complex periodic functions, with Fourier coefficients a_n and b_n, respectively. Express $\int_{-\pi}^{\pi} f\bar{g}$ in terms of the a_n and b_n. $\left(\text{Hint: Consider } \int_{-\pi}^{\pi} |f + g|^2.\right)$

3-14. The Fourier series $\Sigma_{-\infty}^{\infty} a_n e^{in\theta}$ is called *absolutely convergent* if $\Sigma_{-\infty}^{\infty} |a_n| < \infty$. Prove that every absolutely convergent Fourier series is uniformly convergent. Which of the following functions (defined for $|\theta| < \pi$) have absolutely convergent Fourier series?

(i) $f(\theta) = \theta$.

(ii) $f(\theta) = \theta^2$.

(iii) $f(\theta) = 1$ for $\theta > 0$, $f(\theta) = -1$ for $\theta < 0$.

(iv) $f(\theta) = |\theta|$.

3-15. Suppose that $f(\theta)$ is a continuous periodic piecewise differentiable function. Prove that

(i) $f(\theta) = f(0) + \int_0^\theta g(t)\, dt$ for a piecewise continuous g.

(ii) $\int_{-\pi}^{\pi} g = 0$.

(iii) If the Fourier coefficients of f and g are a_n and b_n respectively, then $a_n = b_n/in$ for $n \neq 0$.

(iv) The Fourier series of f is absolutely convergent. (See Exercises 3-4, 3-9, and 3-14.)

3-16. A function f is called *Lipshitz continuous* if there is a constant M such that

$$|f(\theta) - f(\varphi)| \leq M |\theta - \varphi| \tag{3-19}$$

for all θ and φ. Prove the following.

(i) If f has period 2π and satisfies (3-19) and

$$u(r,\, \theta) = \int_{-\pi}^{\pi} P(r,\, \varphi) f(\theta - \varphi)\, d\varphi$$

$$= \int_{-\pi}^{\pi} P(r,\, \theta - \varphi) f(\varphi)\, d\varphi$$

then $\partial u/\partial\theta$ exists for $r < 1$ and

$$\left| \frac{\partial u}{\partial \theta} \right| \leq M.$$

(ii) If f satisfies (3-19), then

$$\sum_{-\infty}^{\infty} |a_n| \leq |a_0| + 2M \sqrt{\sum_1^\infty \frac{1}{n^2}}.$$

In particular, the Fourier series of f is absolutely convergent.

3-17. Let f be a continuous function with Fourier coefficients a_n.

(i) If f has continuous derivatives of order m, then $\Sigma n^{2m} |a_n|^2 < \infty$, and $n^m a_n \to 0$.

(ii) If $|n^m a_n|$ has a bound independent of n, then f has continuous derivatives of order $m - 2$.

3-18.   (Legendre polynomials).

(i)   Prove that there is for each integer $n \geq 0$ a unique polynomial $P_n(x) = \Sigma_0^n a_j x^j$ satisfying $P_n(1) = 1$ and $\int_{-1}^1 P_n(x) x^m \, dx = 0$ for $0 \leq m < n$. (The P_n are called the *Legendre polynomials.*)

(ii)   Prove that any polynomial Q of degree $\leq N$ can be written in the form $Q = \Sigma_0^N a_n P_n$ for some constant a_n.

(iii)   If f is a continuous function on $[-1, 1]$, define its Legendre coefficients by

$$L_n = \frac{\int_{-1}^1 f(x) \, P_n(x) \, dx}{\|P_n\|^2},$$

where

$$\|P_n\|^2 = \int_{-1}^1 P_n(x)^2 \, dx.$$

Prove the following analog of Theorem 3-3.

(a)   If Q is any polynomial of degree $\leq N$, then $\int_{-1}^1 |f - Q|^2 > \int_{-1}^1 |f - \Sigma_0^N L_n P_n|^2$ unless $Q = \Sigma_0^N L_n P_n$.

(b) $\int_{-1}^1 |f - \Sigma_0^N L_n P_n|^2 \to 0$ as $N \to \infty$.

(c) $\Sigma_0^\infty |L_n|^2 \|P_n\|^2 = \int_{-1}^1 |f|^2$.

3-19.   For n and m integers let $e_{nm}(x, y) = e^{inx+imy}$, x and y real. For functions f and g continuous on the rectangle $R: |x| \leq \pi$, $|y| \leq \pi$, let

$$(f, g) = \frac{1}{4\pi^2} \int_{-\pi}^\pi \int_{-\pi}^\pi f(x, y) \bar{g}(x, y) \, dx \, dy.$$

(i)   Prove that $(e_{nm}, e_{jk}) = 0$ unless $n = j$ and $m = k$.

(ii)   Prove that if f is continuous on R and $f(-\pi, y) = f(\pi, y)$ for all y, $f(x, -\pi) = f(x, \pi)$ for all x, then

$$\sum_{n=-\infty}^\infty \sum_{m=-\infty}^\infty r^{|n|+|m|}(f, e_{nm}) e_{nm}$$

converges uniformly to f as $r \to 1-$. (This step may be replaced by Exercise 3-20.)

(iii)   Prove the analogs of Theorems 3-1, 3-2, and 3-3 that follow from part (ii).

3-20.   (Trigonometric approximation for functions of two variables).

(i)   Show that the set $\{(x, y): \delta + \cos x \cos y \geq 1\}$ can be made to lie arbitrarily close to the set of lattice points $\{(x, y): x = 2\pi n, y = 2\pi m, n \text{ and } m \text{ integers}\}$, by choosing δ sufficiently small.

(ii)   Let

$$K_{\delta,n}(x, y) = \frac{(\delta + \cos x \cos y)^{2n}}{\int_{-\pi}^\pi \int_{-\pi}^\pi (\delta + \cos x \cos y)^{2n} \, dx \, dy}.$$

Show that

$$\int_{-\pi}^{\pi} \int_{-\pi}^{\pi} K_{\delta,n}(x, y) \, dx \, dy = 1.$$

Given $\epsilon > 0$ and $\eta > 0$, then δ and n can be chosen so that $|K_{\delta,n}(x, y)| < \epsilon$ in the set

$$\{(x, y): \eta < |x| \leq \pi, \eta < |y| \leq \pi\}.$$

(iii)   Prove that, given an f continuous and doubly periodic (that is, $f(x + 2\pi, y) = f(x, y + 2\pi) = f(x, y)$), and given $\epsilon > 0$, then δ and n can be chosen so that

$$\left| f(x', y') - \int_{-\pi}^{\pi} \int_{-\pi}^{\pi} K_{\delta,n}(x, y) f(x' - x, y' - y) \, dx \, dy \right| < \epsilon$$

for all (x', y').

(iv)   Conclude that the function f in part (iii) can be uniformly approximated by a trigonometric polynomial of the form

$$\sum_{-N}^{N} \sum_{-M}^{M} a_{nm} e^{inx} e^{imy}.$$

4

Fourier Transforms

Fourier transforms were developed after Fourier series, but they have become more significant in both pure and applied mathematics. There are several related reasons for this. One is that Fourier transforms are more convenient for calculation; another, that Fourier series are adapted only to periodic functions on the line, while the transform theory applies (with suitable interpretation) to much more general classes of functions. The suitable interpretation is provided by the theory of distributions, which goes beyond the scope of this book.

Fourier transforms provide the representation of a function f not as a series, but as an integral

$$f(x) = \int_{-\infty}^{\infty} g(\xi)e^{i\xi x}\, d\xi$$

and the main difficulty in the discussion lies in this integration over an infinite interval. Accordingly, such "improper" integrals form the subject of Section 4-1. We hope the reader will notice the close connection between the results of this section and certain theorems on infinite series. The next two sections consider the problem of steady state temperatures in a half plane, as a motivation for the introduction of Fourier integrals.   The further discussion of this problem entails a detour into harmonic functions, to obtain a theorem on uniqueness.   This section, 4-4, may be skipped without any serious loss of continuity, but it contains some more applications of the ideas of Chapter 1.   Sections 4-5 to 4-7 present the various formulas relating to inversion, differentiation, mean square integrals, and multiplication that make Fourier transforms useful, and finally Section 4-8 applies these to a discussion of time-dependent temperature distributions in a rod.

In several places the assumptions are made unnecessarily strong, in order to present a relatively simple proof of an important general principle.   Really satisfactory statements of many of these results rely on· the Lebesgue integral or the theory of distributions.   In view of this it seems wiser not to strain at the maximum generality with the tools at our disposal.   Even so, the reader will find the analysis much denser than in the first three chapters, and might do well to skip the longer proofs on a first reading.

4-1. IMPROPER INTEGRALS

Suppose f is a piecewise continuous function on the real line; by this we mean that in each finite interval $A \leq x \leq B$ the function has at most finitely many points a_j of discontinuity, and that at each such point the one-sided limits $f(a_j+)$ and $f(a_j-)$ exist.   Then

$$F(B) = \int_A^B f$$

exists; if $\lim_{B \to \infty} F(B)$ exists we denote it by

$$\int_A^\infty f \quad \text{or} \quad \int_A^\infty f(x) \, dx$$

and say that f has an improper integral on the interval (A, ∞), or simply that $\int_A^\infty f$ converges.   For example, of the integrals

$$\int_0^\infty \frac{dx}{1 + x^2} \quad \int_1^\infty \frac{dx}{x} \quad \int_0^\infty \sin x \, dx$$

the first converges and the others do not. Similarly, $\int_{-\infty}^{B} f$ converges when $\lim_{A \to -\infty} \int_{A}^{B} f$ exists; and $\int_{-\infty}^{\infty} f$ converges when both $\int_{-\infty}^{0} f$ and $\int_{0}^{\infty} f$ do so. As with infinite series, we say that $\int_{A}^{\infty} f$ converges *absolutely* if $\int_{A}^{\infty} |f|$ converges. The most elementary facts about convergent and absolutely convergent improper integrals are summed up in the next two exercises. The proofs resemble those of corresponding results for infinite series.

Exercises. **4-1.** Let f and g be piecewise continuous on the whole real line.

(i) If $\int_{a}^{\infty} f$ and $\int_{a}^{\infty} g$ exist, and c is any number, then $\int_{a}^{\infty} (f + cg)$ exists and equals $\int_{a}^{\infty} f + c \int_{a}^{\infty} g$.

(ii) If $f = u + iv$ with u and v real, then $\int_{a}^{\infty} f$ exists if and only if $\int_{a}^{\infty} u$ and $\int_{a}^{\infty} v$ exist.

(iii) If $\int_{a}^{\infty} f$ exists for some finite a, then $\int_{b}^{\infty} f$ exists for all finite b, and $\int_{a}^{\infty} f = \int_{a}^{b} f + \int_{b}^{\infty} f$.

(iv) If $\int_{1}^{\infty} f$ exists, then

$$\int_{a}^{\infty} f(y) \, dy = a \int_{1}^{\infty} f(ax) \, dx.$$

(v) At any point where f is continuous, the function $F(x) = \int_{x}^{\infty} f$ has the derivative

$$F'(x) = -f(x).$$

4-2. Let f and g be piecewise continuous on the real line.

(i) If $f \geq 0$ and $\int_{a}^{B} f$ has a bound independent of B, then $\int_{a}^{\infty} f$ converges to $\sup_{B} \int_{a}^{B} f$.

(ii) If $|f| \leq g$ and $\int_{a}^{\infty} g$ converges, then $\int_{a}^{\infty} f$ converges. (Hint: From part (ii) of the previous exercise, it is enough to consider f real; then let $f = f_{+} - f_{-}$, where $f_{+}(x) = \max(f(x), 0)$ and $f_{-}(x) = -\min(f(x), 0)$, and apply part (i) of this and the previous exercise. Alternatively, a proof may be based on the Cauchy convergence criterion.)

(iii)   If $\int_a^\infty |f|$ converges, then $\int_a^\infty f$ converges.

(iv) $\int_{-\infty}^\infty f$ converges absolutely if $|f(x)| \leq c(1 + |x|)^{-1-\epsilon}$ for some constants c and $\epsilon > 0$.

(v)   If $\int_a^\infty f$ converges, then $\lim_{b \to +\infty} \int_b^\infty f = 0$.

After the above facts, the most useful are the analogs of integration and differentiation of series term by term, presented in the following proposition. The basic assumption here is that of *uniform convergence*. Let f be a continuous function of two real variables. We say that $\int_a^\infty f(x, y)\, dx$ converges uniformly for $b \leq y \leq c$ if the integral converges for each such y, and given $\epsilon > 0$ there is a number A such that

$$\left| \int_a^B f(x, y)\, dx - \int_a^\infty f(x, y)\, dx \right| < \epsilon$$

for all $B > A$ and all y in $b \leq y \leq c$.

Theorem 4-1.   *Let f be continuous in the region $a \leq x < \infty$, $b \leq y \leq c$.*

(i)   *If $\int_a^\infty f(x, y)\, dx$ converges uniformly for $b \leq y \leq c$, then the integral is a continuous function of y on that interval.*

(ii)   *With the hypotheses of (i), $\int_a^\infty \left[\int_b^c f(x, y)\, dy \right] dx$ converges and equals $\int_b^c \left[\int_a^\infty f(x, y)\, dx \right] dy$.*

(iii)   *If $f = \partial F/\partial y$ is continuous, if $\int_a^\infty F(x, y)\, dx$ converges, and $\int_a^\infty f(x, y)\, dx$ converges uniformly for $b \leq y \leq c$, then $\int_a^\infty F(x, y)\, dx$ has the derivative $\int_a^\infty f(x, y)\, dx$ for $b < y < c$.*

(iv)   *If f is continuous for $x \geq a$ and $y \geq b$, and*

$$\int_a^\infty |f(x, y)|\, dx$$

and $\int_b^\infty |f(x, y)|\, dy$ converge uniformly on each finite interval, and one of

$$\int_b^\infty \left[\int_a^\infty |f(x, y)|\, dx \right] dy \qquad \int_a^\infty \left[\int_b^\infty |f(x, y)|\, dy \right] dx$$

converges, then

$$\int_b^\infty \left[\int_a^\infty f(x, y)\, dx \right] dy = \int_a^\infty \left[\int_b^\infty f(x, y)\, dy \right] dx. \quad (4\text{-}1)$$

The proofs, again, resemble those for series. For part (i), recall that $\int_a^A f(x, y)\, dx$ is a continuous function of y for each A. Given y_0 and $\epsilon > 0$, choose A so that $\left| \int_A^\infty f(x, y)\, dx \right| < \epsilon/3$; then choose δ so that $\left| \int_a^A [f(x, y) - f(x, y_0)]\, dx \right| < \epsilon/3$ for $|y - y_0| < \delta$, so that for $|y - y_0| < \delta$ we have

$$\left| \int_a^\infty f(x, y)\, dx - \int_a^\infty f(x, y_0)\, dx \right|$$

$$\leq \left| \int_A^\infty f(x, y)\, dx \right| + \left| \int_a^A f(x, y) - f(x, y_0)\, dx \right|$$

$$+ \left| \int_A^\infty f(x, y_0)\, dx \right| < \epsilon.$$

Concerning part (ii), we know that $\int_b^c \left[\int_a^\infty f(x, y)\, dx \right] dy$ exists, since the function in brackets is continuous. Then

$$\left| \int_a^A \left[\int_b^c f(x, y)\, dy \right] dx - \int_b^c \left[\int_a^\infty f(x, y)\, dx \right] dy \right|$$

$$= \left| \int_b^c \left[\int_a^A f(x, y)\, dx \right] dy - \int_b^c \left[\int_a^\infty f(x, y)\, dx \right] dy \right|$$

$$(4\text{-}2)$$

since we may change the order of a repeated integral over finite intervals. Now given $\epsilon > 0$, choose A_0 so that

$$\left| \int_A^\infty f(x, y)\, dx \right| < \frac{\epsilon}{c - b} \qquad \text{for all } A > A_0$$

and $b \leq y \leq c$. Then the expression in (4-2) is $< \epsilon$ for all $A > A_0$, which proves part (ii).

For part (iii), we have

$$\int_c^y \int_a^\infty f(x, t)\, dx\, dt = \int_a^\infty \int_c^y f(x, t)\, dt\, dx$$

$$= \int_a^\infty F(x, y)\, dx - \int_a^\infty F(x, c)\, dx.$$

By the fundamental theorem of calculus, then, $\int_a^\infty F(x, y)\, dx$ has as derivative $\int_a^\infty f(x, y)\, dx$.

For part (iv) take first the case that $f \geq 0$, and suppose

$$I_1 = \int_b^\infty \int_a^\infty f(x, y)\, dx\, dy$$

is convergent. Then by virtue of part (ii),

$$I_2 = \lim_{A \to \infty} \int_a^A \int_b^\infty f(x, y)\, dy\, dx$$

$$= \lim_{A \to \infty} \int_b^\infty \int_a^A f(x, y)\, dx\, dy \leq I_1.$$

Now reversing the argument yields $I_1 \leq I_2$, and (4-1) is established.

In the general case write

$$f = f_1 - f_2 + if_3 - if_4$$

with $0 \leq f_j \leq |f|$, by taking, for instance, $f_1 = \max(0, \text{Real part of } f)$. Then the hypotheses of part (iv) apply to each f_j, and (4-2) follows from what we have shown and Exercise 4-2(i).

Theorem 4-1 has surprising applications to the evaluation of several of the integrals that arise in connection with Fourier transforms. One of the most important is

$$\int_0^\infty \frac{\sin Ax}{x}\, dx = \frac{\pi}{2} \qquad (A \geq 0). \tag{4-3}$$

This can be derived more or less directly from the fact that the integral of the Dirichlet kernel of Section 1-8 is 1:

$$\int_{-\pi}^\pi D_n(\theta)\, d\theta = 1.$$

A slightly simpler derivation is obtained by complicating the integral, and considering

$$I(y) = \int_0^\infty e^{-xy} \frac{\sin Ax}{x}\, dx \qquad (y \geq 0). \tag{4-4}$$

Then

$$I'(y) = -\int_0^\infty e^{-xy} \sin Ax\, dx \qquad (y > 0)$$

since the differentiated integral converges uniformly in every interval $y \geq \epsilon$. The integral for I' is the imaginary part of

$$- \int_0^\infty e^{-xy+iAx} \, dx = - \frac{y + iA}{y^2 + A^2}$$

that is,

$$I'(y) = - \frac{A}{y^2 + A^2}$$

so

$$I(y) = c - \text{arc tan} \frac{y}{A} \qquad (y > 0) \qquad (4\text{-}5)$$

for some constant c. The evaluation is completed by showing that $I(y)$ is continuous for $y \geq 0$, and converges to zero as $y \to +\infty$. From the second fact we find the constant c in (4-5) is $\pi/2$, and from the first we have

$$I(0) = \lim_{y \to 0} \left(\frac{\pi}{2} - \text{arc tan} \frac{y}{A} \right) = \frac{\pi}{2}$$

which is (4-3).

The fact that $\lim_{y \to \infty} I(y) = 0$ is straightforward, since

$$|I(y)| \leq \int_0^\infty e^{-xy} \, dx = \frac{1}{y}.$$

The continuity follows from the uniform convergence of the integral, according to Theorem 4-1. This uniform convergence can be demonstrated by appealing to the alternating series

$$\sum_0^\infty \int_{n\pi/A}^{(n+1)\pi/A} e^{-xy} \frac{\sin Ax \, dx}{x}. \qquad (4\text{-}6)$$

The series (4-6) converges since the absolute values of the successive terms decrease monotonically to zero, and the signs of the terms alternate. To see that $I(y)$ converges uniformly to the sum of (4-6), let $\epsilon > 0$ be given, and take an integer

$N > 2/\epsilon$. Then for $a > N\pi/A$ we have

$$\left| \int_0^a e^{-xy} \frac{\sin Ax}{x}\, dx - \sum_0^\infty \int_{n\pi/A}^{(n+1)\pi/A} e^{-xy} \frac{\sin Ax}{x}\, dx \right|$$

$$= \left| \int_a^{m\pi/A} e^{-xy} \frac{\sin Ax}{x}\, dx \right.$$

$$\left. - \sum_m^\infty \int_{n\pi/A}^{(n+1)\pi/A} e^{-xy} \frac{\sin Ax}{x}\, dx \right|, \quad (4\text{-}7)$$

where m is chosen so that $(m - 1)\pi/A < a \le m\pi/A$.

The first integral on the right is dominated by

$$\frac{(m\pi/A) - a}{a} \le \frac{\pi}{Aa} \le \frac{\pi}{N\pi} < \frac{\epsilon}{2}$$

and the alternating series on the right of (4-7) is dominated by its first term, which is likewise $< \epsilon/2$. Since the restriction $a \ge N\pi/A$ is independent of y, the integral (4-4) converges uniformly, and the derivation is complete.

Exercises. **4-3.** The value of $\int_0^\infty (\sin Ax)/x\, dx$ can be found by replacing $(\sin Ax)/x$ by $\int_0^\infty e^{-xy} \sin Ax\, dy$ and changing the order of integration. Can you justify this?

4-4. For some constant c,

$$I(y) = \int_0^\infty \exp(-a^2x^2) \cos xy\, dx = c \exp\left(-\frac{y^2}{4a^2}\right).$$

(Hint: Show that $I' = -yI/2a^2$. The constant c can be obtained by a trick that evaluates $I(0)$; we shall obtain it later from the Fourier inversion theorem.)

4-5. Show that $\int_0^\infty \sin(x^2)\, dx$ converges, by changing the variable to $t = x^2$. Thus, in contrast to series, $\int_0^\infty f(x)\, dx$ can converge without $\lim_{x\to\infty} f(x) = 0$.

4-6. Construct a function $f \ge 0$ with $\int_0^\infty f$ convergent, but not $f(x) \to 0$ as $x \to \infty$.

4-7. The difficulties faced in part (iv) of Theorem 4-1 are not imaginary, since

$$\int_1^\infty \left[\int_1^\infty \frac{x - y}{(x + y)^3}\, dx \right] dy \ne \int_1^\infty \left[\int_1^\infty \frac{x - y}{(x + y)^3}\, dy \right] dx.$$

4-2. THE DIRICHLET PROBLEM IN A
HALF PLANE

Suppose there is a bounded, continuous steady state temperature distribution in an infinitely large plate, and the temperature distribution along one edge is known. Do these conditions determine the temperature in the plate? The mathematical formulation of this question is Dirichlet's problem in a half plane: do the conditions

v continuous for $y \geq 0$, $-\infty < x < \infty$ and

v has second derivatives for $y > 0$

$$\frac{\partial^2 v}{\partial x^2} + \frac{\partial^2 v}{\partial y^2} = 0 \qquad \text{for} \quad -\infty < x < \infty, 0 < y \leq \infty \quad (4\text{-}8)$$

$$v(x, 0) = g(x) \qquad \text{for} \quad -\infty < x < \infty$$

$$g \text{ bounded and continuous} \quad (4\text{-}9)$$

$$v \text{ bounded for } -\infty < x < \infty, y \geq 0 \qquad (4\text{-}10)$$

determine a unique function v? Recall that condition (4-8) is the equation of a steady state temperature distribution in a plate, found in Exercise 1-1.

In order to find a solution, we attempt again (as in the corresponding problem in a disk) to form an appropriate linear combination of product solutions, each of which satisfies the linear homogeneous conditions (4-8) and (4-10). Setting $v = X(x)Y(y)$ in (4-8) yields, after separation,

$$-\frac{X''}{X} = \frac{Y''}{Y} = c.$$

The solutions are, for $c \neq 0$,

$$X = a \exp(i \sqrt{c}\, x) + b \exp(-i \sqrt{c}\, x)$$

$$Y = A \exp(\sqrt{c}\, y) + B \exp(-\sqrt{c}\, y).$$

Imposing condition (4-10), and requiring $XY \neq 0$, yields $c > 0$ because of the form of the solution for X, and then $A = 0$ in the solution for Y. Absorbing the constant B into *a* and *b* leaves us with the solutions

$$\exp(-\sqrt{c}\,y)(a\exp(i\sqrt{c}\,x) + b\exp(-i\sqrt{c}\,x)) \qquad (c > 0).$$

The reader can check that the product solution is a constant function when the separation constant $c = 0$. All these solutions can be represented conveniently by introducing a new parameter $\xi = \pm\sqrt{c}$, and writing the product solutions,

$$v_\xi(x, y) = e^{-|\xi|y}e^{i\xi x} \qquad (4\text{-}11)$$

with one such solution for each ξ in $-\infty < \xi < \infty$. These are all the product solutions $v = X(x)Y(y)$ satisfying (4-8) and (4-10).

There is a striking similarity between these functions and those found for the corresponding problem in a disk, namely $r^{|n|}e^{in\theta}$, $r < 1$. In fact, setting $r = e^{-y}$ and $x = \theta$ makes the two sets identical, except that in the case of the disk there is one product solution for each integer, and in the present case there is one for each real number. This is a fundamental difference that frequently occurs in the passage to infinite domains.

In the case of the disk, the product solutions were superimposed in a series

$$u(r, \theta) = \sum_{-\infty}^{\infty} a_n r^{|n|} e^{in\theta}$$

and the question was how to choose a_n so that

$$u(1, \theta) = \sum_{-\infty}^{\infty} a_n e^{in\theta} = f(\theta).$$

The corresponding combination of the solutions (4-11) can reasonably be made with an integral,

$$v(x, y) = \int_{-\infty}^{\infty} a(\xi)e^{-|\xi|y}e^{i\xi x}\,d\xi. \qquad (4\text{-}12)$$

The question of satisfying condition (4-9) becomes how to choose the function a so that

$$g(x) = \int_{-\infty}^{\infty} a(\xi)e^{i\xi x}\,d\xi. \qquad (4\text{-}13)$$

Put in slightly different terms, if the integral in (4-13) is known for each x, how can $a(\xi)$ be determined?

Recall that, in the series case, the formula

$$a_n = \frac{1}{2\pi} \int_{-\pi}^{\pi} f(\theta) e^{-in\theta}\, d\theta$$

was derived by formal (not rigorously justified) integration of the series $\Sigma\, a_n e^{in\theta}$; the result of the illegitimate calculation was used to write down a solution whose validity could be checked under suitable hypotheses. We follow the same course here, in relating the function a in (4-13) to appropriate integrals involving g; the gap between the formal calculations and a rigorous deduction, however, is much larger in the present case. We have

$$\int_{-A}^{A} g(x) e^{-ix\eta}\, dx \,-\, 2\pi a(\eta)$$

$$= \int_{-A}^{A} \int_{-\infty}^{\infty} a(\xi) e^{i(\xi-\eta)x}\, d\xi\, dx \,-\, 2\pi a(\eta)$$

$$= \int_{-\infty}^{\infty} a(\xi) \int_{-A}^{A} e^{i(\xi-\eta)x}\, dx\, d\xi \,-\, 2\pi a(\eta)$$

$$= \int_{-\infty}^{\infty} 2a(\xi)[\sin(\xi-\eta)A](\xi-\eta)^{-1}\, d\xi \,-\, 2\pi a(\eta)$$

$$= 2\int_{-\infty}^{\infty} [a(\xi) - a(\eta)](\xi-\eta)^{-1} \sin A(\xi-\eta)\, d\xi$$

$$= 2\int_{-\infty}^{\infty} [a(\tau+\eta) - a(\eta)]\tau^{-1} \sin A\tau\, d\tau$$

$$\to 0 \qquad \text{as} \quad A \to \infty.$$

The next to last equality above uses the integral (4-3), and the claim that the limit is zero is based on the Riemann lemma, Section 1-8. This is the greatest weakness in the derivation, since nothing has been done to justify the application of the Riemann lemma in such a situation. Nonetheless there are conditions, given in the proof of the Fourier inversion formula below, under which these manipulations are justified. For the moment, we accept them in the same spirit as those leading to the Fourier coefficients, and make the following definition.

Definition. *If g is a piecewise continuous function and $\int_{-\infty}^{\infty} |g|$ converges, then the function*

$$\hat{g}(\eta) = \frac{1}{2\pi} \int_{-\infty}^{\infty} e^{-ix\eta} g(x)\, dx \qquad (4\text{-}14)$$

is the Fourier transform *of g. (Note that, from Exercise (4-2), the integral for $\hat{g}(\eta)$ converges.) The integral*

$$\int_{-\infty}^{\infty} \hat{g}(\xi) e^{i\xi x}\, d\xi \qquad (4\text{-}15)$$

(which may or may not converge) is called the Fourier integral *of g.*

The reader must be warned that there is no single customary definition of the Fourier transform and the Fourier integral. The variations arise in the treatment of the factor $1/2\pi$, and the sign of the exponential. For each definition of the Fourier transform there is only one reasonable definition of the Fourier integral, however. The various definitions have the form

$$\hat{g}(\eta) = a \int_{-\infty}^{\infty} e^{ib\eta x} g(x)\, dx$$

for some constants a and b, with the corresponding Fourier integral being

$$\frac{|b|}{2\pi a} \int_{-\infty}^{\infty} e^{-ib\xi x} \hat{g}(\xi)\, d\xi.$$

We have taken $a = 1/2\pi$, $b = -1$. Other definitions use various combinations of $a = 1$, $a = 1/\sqrt{2}$, $b = \pm 1$, $b = \pm 2\pi$. The factor $|b|/2\pi a$ in the Fourier integral is selected so that the Fourier integral of g is actually equal to g for a general class of functions.

4-3. POISSON'S KERNEL FOR A HALF PLANE

In attempting to solve the problem (4-8)–(4-10) we arrived, by the separation method, at the solution

$$v(x, y) = \int_{-\infty}^{\infty} \hat{g}(\xi) e^{-|\xi| y} e^{i\xi x}\, d\xi$$

with $\hat{g}$ the Fourier transform of g. This corresponds to the formal solution of the Dirichlet problem in a disk found in Section 1-3. In the present section we give the results corresponding to Sections 1-4 and 1-5, and show in particular that the above expression for v actually solves the problem.

Proceeding again without strict regard for the rules, we have, on substituting the definition of $\hat{g}$ and changing the order of integration,

$$v(x, y) = \int_{-\infty}^{\infty} g(t) \left[\frac{1}{2\pi} \int_{-\infty}^{\infty} e^{-|\xi|y} e^{i\xi(x-t)} \, d\xi \right] dt$$

$$= \frac{1}{\pi} \int_{-\infty}^{\infty} g(t) \frac{y}{(x-t)^2 + y^2} \, dt \qquad (4\text{-}16)$$

which is the *Poisson integral formula for a half plane*. Poisson's kernel for a half plane is

$$P(x - t, y) = \frac{1}{\pi} y[(x-t)^2 + y^2]^{-1} \qquad (y > 0). \quad (4\text{-}17)$$

The integral in brackets in (4-16) is evaluated as follows:

$$\int_{-\infty}^{\infty} e^{-|\xi|y} e^{i\xi s} \, d\xi = \int_0^{\infty} e^{(is-y)\xi} \, d\xi$$

$$+ \int_{-\infty}^0 e^{(is+y)\xi} \, d\xi$$

$$= -(is - y)^{-1} + (is + y)^{-1} = 2y(s^2 + y^2)^{-1}.$$

The Poisson integral yields the following solution to the Dirichlet problem for a half plane.

Theorem 4-2. *Let g be a bounded, continuous function on the real line. Then the function v given by (4-16) for $y > 0$ satisfies*

(i) $\dfrac{\partial^2 v}{\partial x^2} + \dfrac{\partial^2 v}{\partial y^2} = 0 \qquad$ for $y > 0$.

(ii) $|v(x, y)| \leq \sup_t |g(t)|$.

(iii) $\lim_{y \to 0} v(x, y) = g(x)$.

The convergence in part (iii) *is uniform on each set $\{x: |x| \leq A\}$, and if we set $v(x, 0) = g(x)$, then v is continuous for $y \geq 0$.*

Proof. The reader can check that

$$\frac{\partial^2 P(x - t, y)}{\partial x^2} + \frac{\partial^2 P(x - t, y)}{\partial y^2} = 0$$

for $y > 0$; thus part (i) is established if we show that the expression (4-16) for v can be differentiated under the integral, or that the differentiated integral converges uniformly. Pick an arbitrary $\delta > 0$, and consider y in the interval $\delta < y \leq 1/\delta$; if the differentiations are valid in this interval for every $\delta > 0$, then they are valid for all $y > 0$. The integral for v, differentiated formally once with respect to y, is $\int_{-\infty}^{\infty} h(x, t, y) \, dt$, with

$$h(x, t, y) = \frac{1}{\pi} g(t)[(x - t)^2 - y^2][(x - t)^2 + y^2]^{-2}.$$

Since $(a^2 - y^2)(a^2 + y^2)^{-2} \leq (a^2 + y^2)^{-1}$, we have for $y \geq \delta$ that

$$\left| \int_{|t| \geq A} h(x, t, y) \, dt \right| \leq \frac{\sup |g|}{\pi} \int_{|t| \geq A} \frac{dt}{(x - t)^2 + \delta^2},$$

which tends to zero uniformly. This justifies one differentiation, and the others are nearly identical. Hence part (i) is established.

The claim (ii) follows from the fact that $P \geq 0$ and

$$\int_{-\infty}^{\infty} P(x - t, y) \, dt = 1. \tag{4-18}$$

We leave the one-step proof to the reader.

The convergence in part (iii) is established as in Theorem 1-1. Let $\epsilon > 0$ be given, and choose $\delta > 0$ so that

$$|g(x) - g(t)| < \frac{\epsilon}{2} \quad \text{if} \quad |x - t| < \delta. \tag{4-19}$$

Then, from (4-18)

$$g(x) - v(x, y) = I_1 + I_2$$

with

$$I_1 = \int_{|x - t| < \delta} [g(x) - g(t)] P(x - t, y) \, dt$$

$$I_2 = \int_{|x - t| > \delta} [g(x) - g(t)] P(x - t, y) \, dt.$$

We have from (4-19), (4-18), and the fact that $P > 0$

$$|I_1| < \frac{\epsilon}{2} \int_{|x-t|<\delta} P(x - t, y)\, dt < \frac{\epsilon}{2}.$$

Further,

$$|I_2| \leq 2 \sup|g| \int_{|x-t|>\delta} P(x - t, y)\, dt$$

and

$$\int_{|x-t|>\delta} P(x - t, y)\, dt = 1 - \frac{2}{\pi} \text{ arc tan } \frac{\delta}{y}$$

which is $< \epsilon/2$ for y sufficiently small.

For the uniform convergence, note that a single number δ in (4-19) can be chosen for all $|x| \leq A$, since g is uniformly continuous on any finite interval; thus the estimates of I_1 and I_2 can be made independent of x for $|x| \leq A$.

The final assertion of the theorem is that $v(x, y)$ tends to $g(x_0)$ as (x, y) tends to $(x_0, 0)$. This follows from the uniform convergence. For given $\epsilon > 0$ we may choose $\delta_1 > 0$ so that

$$|v(x, y) - g(x)| < \frac{\epsilon}{2} \qquad \text{for} \quad |x - x_0| < 1 \quad \text{and} \quad 0 < y < \delta;$$

and we may choose $1 > \delta_2 > 0$ so that

$$|g(x) - g(x_0)| < \frac{\epsilon}{2} \qquad \text{for} \quad |x - x_0| < \delta_2.$$

Then for $|x - x_0| < \delta_2$ and $0 < y < \delta_1$,

$$|v(x, y) - g(x_0)| \leq |v(x, y) - g(x)| + |g(x) - g(x_0)| < \epsilon.$$

Exercises. 4-8. Let $g(t)$ be constant, and find $v(x, y)$ in (4-16).

4-9. What modifications are necessary in Theorem 4-2 if g is assumed merely piecewise continuous and bounded?

4-10. Show that $v(x, y) \to 0$ uniformly as $y \to \infty$ if $\int_{-\infty}^{\infty} |g|$ converges. Do the same for the case in which $\int_{-\infty}^{\infty} |g|^2$ converges.

4-11. Show that the v of (4-16) satisfies inf $g \leq v \leq$ sup g, when g is real-valued.

4-12. Suppose the derivative g' is continuous and bounded. Show that $\partial v/\partial x \to g'$. (Hint: Either integrate by parts, or rewrite v in the form

$$\int_{-\infty}^{\infty} g(x - t) P(t, y)\, dt.)$$

4-13. Let g be bounded and continuous, and v be given by (4-16). Suppose for some x_0, that

$$\lim_{y \to \infty} v(x_0, y) = L$$

exists. Prove that for all x, $v(x, y) \to L$ as $y \to \infty$. (When this limit exists, it represents in some sense the mean value of g.)

4-14. Prove the following analog of Theorem 1-2. If f and g are continuous, $\int_{-\infty}^{\infty} |f|$ and $\int_{-\infty}^{\infty} |g|$ converge, and $\hat{f} = \hat{g}$, then $f = g$. (Hint: Use part (iv) of Theorem 4-1 to show that

$$\int_{-\infty}^{\infty} e^{-|\xi| y} \hat{g}(\xi) \, d\xi = \int_{-\infty}^{\infty} g(t) P(x - t, y) \, dt$$

and apply Theorem 4-2.)

4-4. THE MAXIMUM PRINCIPLE FOR HARMONIC FUNCTIONS, AND THE QUESTION OF UNIQUENESS

This section deals not so much with Fourier transforms as with harmonic functions, that is, solutions v of Laplace's equation (4-8). The material is presented as a means of showing that the Poisson integral (4-16) gives the *only* solution of the Dirichlet problem (4-8)–(4-10). This question is much more difficult than its analog in the disk, because the fact that g is bounded and continuous does not guarantee that $\hat{g}$ exists. Moreover, the question whether g equals its Fourier integral is still more difficult.

Fortunately, the discussion of the Dirichlet problem in a disk has provided an important *maximum principle for harmonic functions* (see Exercise 1-16), which we here state, prove, and apply. The formulation we give differs slightly from Exercise 1-16. The statement involves bounded open sets and their closures, but the reader unfamiliar with these ideas may restrict himself to the sets

$$S_R = \{(x, y) : x^2 + y^2 < R^2, y > 0\}. \qquad (4\text{-}20)$$

S_R is *bounded* and *open*. Its boundary is

$$B_R = \{(x, y) : x^2 + y^2 = R^2 \text{ and } y > 0, \text{ or}$$
$$y = 0 \text{ and } |x| < R\}.$$

The closure $\tilde{S}_R$ of S_R is obtained by adjoining the boundary,

$$\tilde{S}_R = \{(x, y): x^2 + y^2 \leq R^2, y \geq 0\}.$$

Theorem 4-3. *Let S be a bounded open set in the plane, with boundary B and closure $\tilde{S}$. If u is a function continuous in $\tilde{S}$ and harmonic in S, then the maximum value of $|u|$ is achieved on the boundary B.*

Proof. Since $|u|$ is a continuous function on the closed bounded set $\tilde{S}$, there is a point (x_0, y_0) in $\tilde{S}$ where $|u|$ achieves its maximum value. If (x_0, y_0) is in B there is nothing to prove. If (x_0, y_0) is in S, let R be the distance from (x_0, y_0) to B. Then the disk

$$D = \{(x - x_0)^2 + (y - y_0)^2 \leq R^2\}$$

lies in $\tilde{S}$ and intersects the boundary B in some point (x_1, y_1); we will show that $|u(x_1, y_1)| = |u(x_0, y_0)|$. Because $|u(x_0, y_0)|$ is a maximum, it is enough to show that $|u(x_1, y_1)| < |u(x_0, y_0)|$ is impossible. From Exercise 1-17,

$$u(x_0, y_0) = \frac{1}{2\pi} \int_{-\pi}^{\pi} u(x_0 + R \cos \theta, y_0 + R \sin \theta)\, d\theta; \quad (4\text{-}21)$$

that is, $u(x_0, y_0)$ is the average of all the values of u on the boundary of the circle of radius R about (x_0, y_0). Because (x_0, y_0) gives a maximum, these boundary values are never larger than $|u(x_0, y_0)|$; and if some are smaller, then the average will have an absolute value $< |u(x_0, y_0)|$, which contradicts (4-21). To obtain this contradiction more formally, let

$$(x_1, y_1) = (x_0 + R \cos \theta_1, x_0 + R \sin \theta_1).$$

If

$$|u(x_0, y_0)| - |u(x_1, y_1)| = d > 0$$

then for some $\delta > 0$ we have

$$|u(x_0 + R \cos \theta, y_0 + R \sin \theta)| < |u(x_0, y_0)| - d/2$$
$$\text{for } |\theta - \theta_1| < \delta \quad (4\text{-}22)$$

while

$$|u(x_0 + R \cos \theta, y_0 + R \sin \theta)| \leq |u(x_0, y_0)| \quad \text{for all } \theta.$$
$$(4\text{-}23)$$

Then (4-21), (4-22), and (4-23) yield the contradiction

$$
\begin{aligned}
|u(x_0, y_0)| &= \left| \frac{1}{2\pi} \int_{|\theta - \theta_1| < \delta} u(x_0 + R \cos \theta, \, y_0 + R \sin \theta) \, d\theta \right. \\
&\quad \left. + \frac{1}{2\pi} \int_{|\theta - \theta_1| > \delta} u(x_0 + R \cos \theta, \, y_0 + R \sin \theta) \, d\theta \right| \\
&\leq \frac{2\delta}{2\pi} \left(|u(x_0, y_0)| - \frac{d}{2} \right) \\
&\quad + (2\pi - 2\delta)(2\pi)^{-1} |u(x_0, y_0)| \\
&= |u(x_0, y_0)| - \frac{\delta d}{2\pi} < |u(x_0, y_0|.
\end{aligned}
$$

The form of the maximum principle given in Exercise 1-16, is stronger than Theorem 4-3; there it is claimed that if $|u|$ achieves its maximum at a point in S, then u is constant. The proof is quite similar, nonetheless.

One of the most important consequences of the maximum principle is the following uniqueness theorem for the Dirichlet problem on a bounded open set.

Corollary 1. *Let S be a bounded open set with boundary B and closure $\bar{S}$. If u and v are harmonic in S, continuous in $\bar{S}$, and $u = v$ on B, then $u = v$ in all of $\bar{S}$.*

Proof. The function $u - v$ is continuous in $\bar{S}$, harmonic in S, and vanishes on B. By the maximum principle, $|u - v| \leq 0$.

The point of this result is that a solution of the Dirichlet problem, obtained by any method, is *the* solution. This is exploited in the next result, where we solve a Dirichlet problem on the set S_R of (4-20).

Corollary 2. *Let u be continuous on $\bar{S}_R$ and harmonic in S_R, and suppose $u(x, 0) = 0$ for $|x| \leq R$. Then for $r < R$, $0 \leq \theta \leq \pi$,*

$$
u(r \cos \theta, \, r \sin \theta)
$$

$$
= \frac{2}{\pi} \int_0^\pi \frac{\rho(1 - \rho^2) \sin \theta \sin t \, u(R \cos t, \, R \sin t) \, dt}{(1 - 2\rho \cos(\theta - t) + \rho^2)(1 - 2\rho \cos(\theta + t) + \rho^2)}
$$

$$
\rho = \frac{r}{R}. \tag{4-24}
$$

Proof. By Corollary 1 on uniqueness, it is enough to show that the integral in (4-24) is a harmonic function with the same boundary values as u. Such a function can be constructed out of the Poisson integral (1-10) by using an appropriate symmetry. Let

$$f(\theta) = u(R \cos \theta, R \sin \theta) \qquad \text{for} \quad 0 \leq \theta \leq \pi$$

and define

$$f(\theta) = -u(R \cos \theta, -R \sin \theta) \qquad \text{for} \quad -\pi \leq \theta < 0.$$

Then f is continuous for $-\pi \leq \theta \leq \pi$, and $f(-\pi) = f(\pi) \, (= 0)$. Now consider the temperature distribution v in the disk $x^2 + y^2 < R^2$ with boundary values $v(R \cos \theta, R \sin \theta) = f(\theta)$. Since the values of f below the x axis are the negatives of those above that axis, we expect that the resulting temperature distribution v will vanish on the x axis, hence agree with u on the boundary of S_R, and hence (by the maximum principle) agree with u in all of S_R. We check this by writing the Poisson integral

$$v(r \cos \theta, r \sin \theta) = \frac{1}{2\pi} \int_{-\pi}^{\pi} \frac{1 - \rho^2}{1 - 2\rho \cos(\theta - t) + \rho^2} f(t) \, dt$$

$$\rho = \frac{r}{R}$$

and setting

$$v(R \cos \theta, R \sin \theta) = f(\theta).$$

Then v is harmonic in $x^2 + y^2 < R^2$, continuous in $x^2 + y^2 \leq R^2$, and $v = u$ on the curved boundary of S_R. (See Exercises 1–18.) Using the definition of f we have

$$v(r \cos \theta, r \sin \theta)$$

$$= \frac{1}{2\pi} \int_0^{\pi} \frac{1 - \rho^2}{1 - 2\rho \cos(\theta - t) + \rho^2} u(R \cos t, R \sin t) \, dt$$

$$- \frac{1}{2\pi} \int_{-\pi}^0 \frac{1 - \rho^2}{1 - 2\rho \cos(\theta - s) + \rho^2} u(R \cos(-s), R \sin(-s)) \, ds.$$

The substitution $s = -t$ in the second integral leads to the integral in (4-24). It is evident from that form that $0 = v(r \cos \theta, r \sin \theta)$ if $\theta = 0$, $\theta = \pi$, or $r = 0$. Hence v does indeed agree with u on the boundary of S_R, so $v = u$ in S_R, and the integral for v is valid for u.

Finally, we use the integral formula of Corollary 2 to prove the uniqueness of solutions of the Dirichlet problem (4-8)–(4-10).

Theorem 4-4. *Let u and v be functions bounded and continuous in the half plane $\{(x, y): y \geq 0\}$, harmonic for $y > 0$, and with $u(x, 0) = v(x, 0)$ for all x. Then $u(x, y) = v(x, y)$ for $y > 0$ also.*

Proof. Let $w = u - v$, so w is bounded, continuous, harmonic for $y > 0$, and $w = 0$ on the x axis. Let $(r \cos \theta, r \sin \theta)$ be any fixed point with $0 < \theta < \pi$, and choose $R > r$. Then from Corollary 2,

$$w(r \cos \theta, r \sin \theta)$$

$$= \frac{2}{\pi} \int_0^\pi \frac{\rho(1 - \rho^2) \sin \theta \sin t \, w(R \cos t, R \sin t) \, dt}{(1 - 2\rho \cos(\theta - t) + \rho^2)(1 - 2\rho \cos(\theta + t) + \rho^2)}$$

$$\rho = \frac{r}{R}. \tag{4-25}$$

We hold r fixed, and let $R \to \infty$, which means $\rho \to 0$. Now in the integral (4-25), $w(R \cos t, R \sin t)$ is bounded by a constant M independent of R, and

$$1 - 2\rho \cos \phi + \rho^2 \geq 1 - 2\rho + \rho^2 = (1 - \rho)^2$$

for any ϕ, so

$$\left| w(r \cos \theta, r \sin \theta) \right| \leq 2M \frac{\rho(1 - \rho^2)}{(1 - \rho)^4} \to 0 \qquad \text{as} \quad \rho \to 0.$$

Thus $w = 0$, that is, $u = v$, and the proof is complete.

There is a simple physical idea here that has probably been hidden by the computations. We consider the function w in S_R as the solution of a Dirichlet problem with boundary values which are bounded on the circular arc, and vanish on the x axis. When R is very large, the most influential boundary values are the zeros. (See Fig. 4-1.)

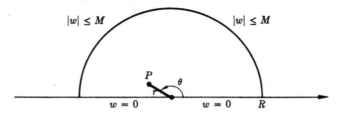

$$P = (r \cos \theta, r \sin \theta), \quad |w(P)| \leq 2M \frac{\rho(1 - \rho^2)}{(1 - \rho)^4}, \quad \rho = r/R.$$

FIGURE 4-1. *Illustration of proof of Theorem 4-4.*

Exercises. **4-15.** Suppose u is continuous in $\{(x, y): y \geq 0\}$, harmonic for $y > 0$, and $u(x, 0) = e^{i\xi x}$, ξ real. Show

(a) $u(x, y) = e^{i\xi x - |\xi|y}$

(b) $u(x, y) = \frac{1}{\pi} \int_{-\infty}^{\infty} e^{it\xi} \frac{y}{(x - t)^2 + y^2} dt.$

(This is an unusual way to evaluate the integral in (b).)

4-16. Find an unbounded function w harmonic in the whole plane and vanishing for $y = 0$.

4-17. If g is a continuous function of period 2π with Fourier coefficients a_n, the solution of the Dirichlet problem (4-8)–(4-10) is

$$v(x, y) = \sum_{-\infty}^{\infty} a_n e^{inx} e^{-|n|y}.$$

4-18. Suppose v is harmonic in S_R, continuous in $\bar{S}_R$, and $v = 0$ on the x axis. Show that the function u defined by

$$u(x, y) = v(x, y) \qquad (y \geq 0)$$
$$u(x, y) = -v(x, -y) \qquad (y < 0)$$

is harmonic in the disk $x^2 + y^2 < R^2$. (See Corollary 2 above. The result of this exercise is known as the *reflection principle* for harmonic functions.)

4-5. FOURIER INVERSION AND THE PLANCHEREL FORMULA

This and the following two sections present most of the basic facts about Fourier transforms. We begin with a result giving conditions under which a function equals its Fourier integral (4-15); such a result is called a *Fourier inversion theorem*. The proof is suggested by the corresponding proof in the case of series (Theorem 1-4), and by the discussion leading up to the definition of the Fourier integral in Section 4-2.

Theorem 4-5. *Suppose g is continuous and piecewise differentiable, that there is a constant M with $|g'(x)| \leq M$ at all points where g' exists, and that $\int_{-\infty}^{\infty} |g|$ converges. Then*

$$|g(x)|^2 \leq M \int_{-\infty}^{\infty} |g| \qquad (4\text{-}26)$$

for all x. Further, if $\hat{g}$ is the Fourier transform of g, then

$$g_A(x) = \int_{-A}^{A} e^{i\xi x} \hat{g}(\xi) \, d\xi \qquad (4\text{-}27)$$

converges to $g(x)$ uniformly on $-\infty < x < \infty$.

Proof. The bound on $|g'|$ implies that

$$|g(x) - g(y)| \leq M|x - y| \qquad (4\text{-}28)$$

for all x and y. This follows from the mean value theorem if g is differentiable in the interval from x to y. In any case, there are only finitely many points $a_1, \ldots, a_{n-1}$ between x and y at which g fails to be differentiable, and setting $a_0 = x$, $a_n = y$, we have

$$|g(x) - g(y)| \leq \sum_0^n |g(a_{j+1}) - g(a_j)|$$

$$\leq M \sum_0^n |a_{j+1} - a_j| = M|x - y|.$$

The condition (4-28), called a *Lipschitz condition*, has a simple graphical interpretation if g is real-valued: it says that if $P = (x, g(x))$ is any point in the graph of g, then the rest of the graph of g lies between the two lines of slope $\pm M$ passing through P. (See Fig. 4-2.) This suggests the proof of the inequality (4-26), which says that

$$\int |g| \geq \text{area of triangle } PP_1P_2 = |g(x)| \cdot \left(\frac{|g(x)|}{M} \right).$$

We leave it to the reader to provide a formal proof.

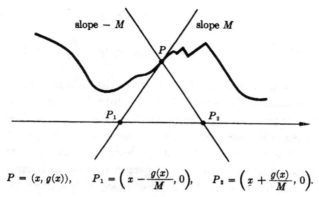

$$P = (x, g(x)), \qquad P_1 = \left(x - \frac{g(x)}{M}, 0 \right), \qquad P_2 = \left(x + \frac{g(x)}{M}, 0 \right).$$

FIGURE 4-2. *Illustration of proof of inequality (4-26).*

To show the uniform convergence of g_A, write

$$g_A(x) = \frac{1}{2\pi} \int_{-A}^{A} \int_{-\infty}^{\infty} e^{i\xi x} e^{-i\xi y} g(y) \, dy \, d\xi$$

$$= \frac{1}{\pi} \int_{-\infty}^{\infty} g(y)(x - y)^{-1} \sin A(x - y) \, dy.$$

Theorem 4-1(ii) justifies the change in order of integration made here. Now given $\epsilon > 0$, let $\delta = \epsilon/8M$, and choose a number B with

$$\int_{|y|>B} |g(y)| \, dy < \frac{\delta\epsilon}{4}. \tag{4-29}$$

Then, since $\int_{-\infty}^{\infty} (x - y)^{-1} \sin A(x - y) \, dy = \pi$, we can write

$$\pi[g_A(x) - g(x)]$$
$$= \int_{|x-y|<\delta} [g(y) - g(x)](x - y)^{-1} \sin A(x - y) \, dy$$
$$- g(x) \int_{|x-y|>\delta} (x - y)^{-1} \sin A(x - y) \, dy$$
$$+ \int_{|y|\le B, |x-y|\ge\delta} g(y)(x - y)^{-1} \sin A(x - y) \, dy$$
$$+ \int_{|y|\ge B, |x-y|\ge\delta} g(y)(x - y)^{-1} \sin A(x - y) \, dy. \tag{4-30}$$

The first term on the right is bounded by $2M\delta = \epsilon/4$, from

(4-28) and the choice of δ. The second is bounded by

$$2 \sup|g| \left| \int_\delta^\infty t^{-1} \sin At \cdot dt \right| = 2 \sup|g| \left| \int_{\delta A}^\infty t^{-1} \sin t \, dt \right|$$

which tends to zero as $A \to \infty$; thus the second term is $< \epsilon/4$ if A is sufficiently large. For the third term on the right of (4-30), suppose temporarily that $-B + \delta < x < B - \delta$, so the integration goes over the two intervals $-B < y < x - \delta$ and $x + \delta < y < B$. We shall show that the integral over the first of these intervals converges to zero uniformly in x. The other interval, and the other possible locations of x, can be handled in exactly the same way. The estimation is a simple integration by parts:

$$\int_{-B_j}^{x-\delta} |g(y)(x - y)^{-1} \sin A(x - y) \, dy$$

$$= -\frac{1}{A} \int_{-B}^{x-\delta} [g(y)(x - y)^{-1}]' \cos A(x - y) \, dy$$

$$+ \frac{g(x - \delta) \cos A\delta}{A\delta} - \frac{g(-B) \cos A(x - B)}{A(x - B)}$$

which converges to zero uniformly for $x < B - \delta$, as $A \to \infty$. Thus the third term is $< \epsilon/4$ when A is sufficiently large. Finally, the fourth term is $< \epsilon/4$ by (4-29). Thus $\pi|g_A(x) - g(x)| < \epsilon$ for A sufficiently large, and the theorem is proved.

Exercise. **4-19.** What is the conclusion of Theorem 4-5 if it is assumed only that g is piecewise differentiable and $\int_{-\infty}^\infty |g|$ converges? Can g_A converge uniformly? (Compare Theorem 1-4.)

One consequence of the inversion theorem is the following analog of Parseval's equality.

Theorem 4-6. *Let g satisfy the hypotheses of Theorem 4-5. Then $\int_{-\infty}^\infty |g|^2$ and $\int_{-\infty}^\infty |\hat{g}|^2$ converge, and*

$$\int_{-\infty}^\infty |g|^2 = 2\pi \int_{-\infty}^\infty |\hat{g}|^2.$$

This result, called the *Plancherel formula*, is quite important for applications both in physics and in mathematics.

Proof. The convergence of $\int_{-\infty}^{\infty} |g|^2$ follows from the fact that

$$|g(x)|^2 \le |g(x)| \max|g|,$$

since g is bounded and $\int_{-\infty}^{\infty} |g(x)|\, dx$ converges.

As for $\int|\hat{g}|^2$, we have

$$\int_{-A}^{A} |\hat{g}(\xi)|^2\, d\xi$$

$$= \frac{1}{4\pi^2} \int_{-A}^{A} \left[\int_{-\infty}^{\infty} g(s)e^{-is\xi}\, ds\right]\left[\int_{-\infty}^{\infty} \bar{g}(t)e^{it\xi}\, dt\right] d\xi.$$

According to Theorem 4-1(ii), we may exchange $\int_{-A}^{A}$ with $\int_{-\infty}^{\infty}$ to obtain

$$\int_{-A}^{A} |\hat{g}(\xi)|^2\, d\xi = \frac{1}{2\pi} \int_{-\infty}^{\infty} \bar{g}(t)g_A(t)\, dt$$

with

$$g_A(t) = \frac{1}{\pi} \int_{-\infty}^{\infty} g(s)(s-t)^{-1} \sin A(s-t)\, ds.$$

Then by Theorem 4-5, $g_A(t)$ converges uniformly to $g(t)$ so

$$\frac{1}{2\pi} \int_{-\infty}^{\infty} |g(t)|^2\, dt - \int_{-A}^{A} |\hat{g}(\xi)|^2\, d\xi$$

$$= \frac{1}{2\pi} \int_{-\infty}^{\infty} \bar{g}(t)[g(t) - g_A(t)]\, dt$$

$$\le \frac{1}{2\pi} \sup|g - g_A| \int_{-\infty}^{\infty} |g| \to 0 \quad \text{as} \quad A \to \infty.$$

The reader may find it surprising that this proof is more direct than the one given for Parseval's equality, Theorem 3-3. The proof above has been simplified by assuming differentiability of g, and by leaving out any discussion of least squares approximation. Some extensions are outlined in the exercises at the end of this section. However, the true value of the

Plancherel formula shows up when Lebesgue integrals are used. In this context, the result can be stated as follows.

Proposition. *If $f \cdot |f|$ is Lebesgue integrable on $(-\infty, \infty)$, then the (Lebesgue) integral*

$$\frac{1}{2\pi} \int_{-A}^{A} f(x) e^{-ix\xi}\, dx$$

exists for each A and ξ, and there is a function $\hat{f}$ (essentially unique) such that $\hat{f} \cdot |\hat{f}|$ is Lebesgue integrable and

$$\int_{-\infty}^{\infty} \left| \frac{1}{2\pi} \int_{-A}^{A} f(x) e^{-ix\xi}\, dx - \hat{f}(\xi) \right|^2 d\xi \to 0 \qquad (4\text{-}31)$$

as $A \to \infty$. Further,

$$\int_{-\infty}^{\infty} \left| \int_{-A}^{A} e^{ix\xi} \hat{f}(\xi)\, d\xi - f(x) \right|^2 dx \to 0$$

as $A \to \infty$, and

$$\int_{-\infty}^{\infty} |f(x)|^2\, dx = 2\pi \int_{-\infty}^{\infty} |\hat{f}(\xi)|^2\, d\xi.$$

Because of the symmetry of these statements, we can deduce that if g is any function with $g \cdot |g|$ Lebesgue integrable, then g is the Fourier transform, in the sense of (4-31), of the function f with

$$\int_{-\infty}^{\infty} \left| \int_{-A}^{A} e^{ix\xi} g(\xi)\, d\xi - f(x) \right|^2 dx \to 0 \qquad \text{as} \quad A \to \infty.$$

To illustrate the power of this formulation, we deduce an analog of the least squares approximation of Section 3-2.

Corollary. *Let $f \cdot |f|$ and $g \cdot |g|$ be Lebesgue integrable on $(-\infty, \infty)$. Then*

$$\int_{-\infty}^{\infty} \left| f(x) - \int_{-A}^{A} g(\xi) e^{i\xi x}\, d\xi \right|^2 dx$$
$$\geq \int_{-\infty}^{\infty} \left| f(x) - \int_{-A}^{A} \hat{f}(\xi) e^{i\xi x}\, d\xi \right|^2 dx.$$

In other words, for any given A, a function of the form

$$\int_{-A}^{A} g(\xi) e^{i\xi x}\, d\xi$$

approximates $f(x)$ best, in the sense of least squares, when $g = \hat{f}$.

Proof. Let

$$h(x) = \int_{-A}^{A} g(\xi) e^{i\xi x} \, d\xi.$$

By the preceding proposition,

$$\hat{h}(\xi) = g(\xi) \qquad (|\xi| < A)$$
$$= 0 \qquad (|\xi| > A)$$

and

$$\int_{-\infty}^{\infty} |f(x) - h(x)|^2 \, dx$$

$$= 2\pi \int_{-\infty}^{\infty} |\hat{f}(\xi) - \hat{h}(\xi)|^2 \, d\xi$$

$$= 2\pi \int_{|\xi| \geq A} |\hat{f}(\xi)|^2 + \int_{|\xi| \leq A} |\hat{f}(\xi) - g(\xi)|^2 \, d\xi.$$

It is clear that this expression is least when $g = \hat{f}$.

Exercises. **4-20.** Compute $\int_{-\infty}^{\infty} e^{it\xi} (\xi^2 + a^2)^{-1} \, d\xi$, with a a constant. (Hint: The Fourier transform of $e^{-a|x|}$ is $(1/\pi) a/(\xi^2 + a^2)$.) Compare this with Exercise 4-15.

4-21. In Exercise 4-4 it was shown that the Fourier transform of $f(x) = \exp(-a^2 x^2)$ was $\hat{f}(\xi) = c \exp(-\xi^2/4a^2)$ for some constant c. Find this constant by using either the inversion theorem or the Plancherel formula.

4-22. (Schwarz's inequality). If f and g are piecewise continuous functions, and $\|f\|^2 = \int_{-\infty}^{\infty} |f|^2$ and $\|g\|^2 = \int_{-\infty}^{\infty} |g|^2$ converge, then $\int_{-\infty}^{\infty} fg$ converges to a value $\leq \|f\| \cdot \|g\|$; moreover, $\int |f + g|^2$ converges, $\|f + g\| \leq \|f\| + \|g\|$, and $|(\|f\| - \|g\|)| \leq \|f - g\|$. (See Section 3-3.)

4-23. (Bessel's inequality).

 (i) Suppose that f is piecewise continuous, and $\int_{-\infty}^{\infty} |f|$ and $\int_{-\infty}^{\infty} |f|^2$ converge. Then there is a sequence $\{f_n\}$ such that each f_n satisfies the hypotheses of Theorem 4-5, $\int_{-\infty}^{\infty} |f_n - f| \to 0$, $\int_{-\infty}^{\infty} |f_n - f|^2 \to 0$, and $\int_{-\infty}^{\infty} |f_n|^2 \to \int_{-\infty}^{\infty} |f|^2$.

(ii)   Use part (i) to show that

$$2\pi \int_{-A}^{A} |f|^2 \le \int_{-\infty}^{\infty} |f|^2$$

for the function f of part (i).

4-24.   (Plancherel formula).   Suppose f and the sequence f_n are as in Exercise 4-23.   Explain why

$$\lim \left| \left(\int_{-\infty}^{\infty} |f|^2 \right)^{1/2} - \left(\int_{-\infty}^{\infty} |f_n|^2 \right)^{1/2} \right| \le \lim \left(\int_{-\infty}^{\infty} |f - f_n|^2 \right)^{1/2}$$

$$\le \frac{1}{2\pi} \lim (\int |f - f_n|^2)^{1/2} = 0.$$

Thus obtain the following generalization of Theorem 4-6: if f is piecewise continuous and $\int_{-\infty}^{\infty} |f|$ and $\int_{-\infty}^{\infty} |f|^2$ are convergent, then

$$2\pi \int_{-\infty}^{\infty} |f|^2 = \int_{-\infty}^{\infty} |f|^2.$$

(See the previous two exercises.)

4-25.   Apply the Plancherel formula (or its generalization in Exercise 4-24) to evaluate $\int_{-\infty}^{\infty} |f|^2$ in the following cases.   Compute f in each case.

   (i) $f(x) = e^{-a|x|}$.
   (ii) $f(x) = 1$ for $a < x < b$, and 0 otherwise.
   (iii) $f(x) = x$ for $|x| < 1$, and 0 otherwise.

4-6. FOURIER TRANSFORMS AND DERIVATIVES

There are two simple formulas relating Fourier transforms and derivatives.

Theorem 4-7.   *If f is piecewise continuous, $\int_{-\infty}^{\infty} |f|$ converges, and $\int_{-\infty}^{\infty} |xf(x)| \, dx$ converges, then $\hat{f}$ has a continuous derivative*

$$\hat{f}'(\xi) = \frac{-i}{2\pi} \int_{-\infty}^{\infty} xf(x)e^{-ix\xi} \, dx. \tag{4-32}$$

If f is continuous and piecewise differentiable, and $\int_{-\infty}^{\infty} |f|$ and $\int_{-\infty}^{\infty} |f'|$ converge, then the transform of f' is

$$\hat{f}'(\xi) = i\xi\hat{f}(\xi). \tag{4-33}$$

Proof. The first statement follows from a differentiation under the integral, which is justified by Theorem 4-1 when f is continuous. Otherwise, let I_j be the intervals in which f is continuous, and differentiate term by term the uniformly convergent series

$$\sum_j \frac{1}{2\pi} \int_{I_j} f(x) e^{-i\xi x} \, dx.$$

The second statement of the theorem comes from an integration by parts:

$$2\pi \widehat{f'}(\xi) = \lim_{A \to \infty} \int_{-A}^{A} e^{-i\xi x} f'(x) \, dx$$

$$= \lim_{A \to \infty} [e^{-i\xi x} f(x)]_{-A}^{A} + i\xi \lim_{A \to \infty} \int_{-A}^{A} e^{-i\xi x} f(x) \, dx.$$

Thus the formula (4-33) is proved if

$$\lim_{x \to \infty} f(x) = 0. \tag{4-34}$$

Suppose (4-34) is false, that is, for some $\epsilon > 0$ there are arbitrarily large values of x with $|f(x)| > \epsilon$. Choose such an x with

$$\int_{|t| > x} |f'(t)| \, dt < \frac{\epsilon}{2}.$$

Suppose that $x > 0$. Then for $t > x$ we have

$$|f(t) - f(x)| = \left| \int_x^t f'(s) \, ds \right| \le \int_x^\infty |f'(s)| \, ds < \frac{\epsilon}{2}$$

so

$$|f(t)| > \frac{\epsilon}{2} \qquad \text{for all} \quad t > x.$$

This contradicts the fact that $\int_{-\infty}^{\infty} |f|$ converges. The same contradiction would arise if x were negative, so the theorem is proved.

To illustrate the use of this result, consider the differential equation

$$u'' - u = g \tag{4-35}$$

with g continuous and $\int_{-\infty}^{\infty} |g|$ convergent. Taking transforms of both sides and using (4-33), we obtain

$$-\xi^2 \hat{u} - \hat{u} = \hat{g}$$

or, if inversion is allowable,

$$u(x) = -\int_{-\infty}^{\infty} e^{i\xi x} \hat{g}(\xi)(1 + \xi^2)^{-1} \, d\xi. \tag{4-36}$$

Actually, *if* $\int_{-\infty}^{\infty} |u|$ is convergent and $u'(x)$ exists for every x, then inversion is allowable, so the formula (4-36) is valid if there is any solution with these properties. In a later section on convolutions we will show that the formula (4-36) defines a function u such that $\int_{-\infty}^{\infty} |u|$ converges, and u and u' are continuous. (See Exercise 4-33.)

Another interesting application of Theorem 4-7 concerns the Hermite functions Φ_n, $n = 0, 1, \ldots$ given by

$$\Phi_0(x) = \exp\left(-\frac{x^2}{2}\right)$$

$$\Phi_{n+1}(x) = x\Phi_n(x) - \Phi_n'(x). \tag{4-37}$$

These functions provide another approach to the Plancherel formula, as suggested by part (vi) of Exercise 4-27 below. Other interesting properties are listed in the same exercise.

Exercises. 4-26. Suppose f is piecewise continuous, and $f(x) = 0$ for $|x| > A$. Prove that $\hat{f}$ has bounded continuous derivatives of all orders. Prove that $\hat{f}$ has a power series

$$\hat{f}(\xi) = \sum_0^{\infty} a_n \xi^n$$

valid on the entire real axis.

4-27. (Hermite functions).

    (i) $\hat{\Phi}_n = i^{-n}\Phi_n/\sqrt{2\pi}$. (See Exercise 4-21 for $\hat{\Phi}_0$.)

    (ii) $\Phi_n = (-1)^n \exp(x^2/2)(d/dx)^n \exp(-x^2)$. (Hint: Show that the functions on the right satisfy the formula (4-37).)

    (iii) $\int_{-\infty}^{\infty} \Phi_n(x)\Phi_m(x) \, dx = 0$ unless $n = m$.

    (iv) $\Phi_n(x) = \exp(-x^2/2)H_n(x)$ for a polynomial H_n of degree n. (These are the Hermite polynomials.)

(v)   If f is continuous and

$$\int_{-\infty}^{\infty} \exp\left(-\frac{x^2}{4}\right) f(x)\, dx$$

converges, and

$$\int_{-\infty}^{\infty} \Phi_n(x) f(x)\, dx = 0 \qquad (4\text{-}38)$$

for all n, then $f = 0$. (Hint: The condition (4-38) shows $\int_{-\infty}^{\infty} x^n$ $\exp(-x^2/2)f(x)\, dx = 0$ for all n; an application of Stirling's formula for $m!$ shows that $\exp(-x^2/4)\Sigma_0^{\infty}(-i\xi x)^m/m!$ converges uniformly to $\exp(-x^2/4)e^{-i\xi x}$, so the Fourier transform of $\exp(-x^2/2)f(x)$ vanishes. Apply Exercise 4-14.)

(vi)   Use parts (i) and (iii) above to show that, if $f = \Sigma_0^N a_n\Phi_n$ for some constants a_n, then

$$\int_{-\infty}^{\infty} |f|^2 = 2\pi \int_{-\infty}^{\infty} |\hat{f}|^2$$

4-7. CONVOLUTION

Recall the integral formulas for the solution of the Dirichlet problem in the circle, (1-10),

$$u(r, \theta) = \frac{1}{2\pi} \int_{-\pi}^{\pi} f(t)\, \frac{1 - r^2}{1 - 2r\cos(\theta - t) + r^2}\, dt$$

and in the half plane, (4-16),

$$v(x, y) = \frac{1}{\pi} \int_{-\infty}^{\infty} g(t)\, \frac{y}{(x - t)^2 + y^2}\, dt.$$

These are two examples of an important way of combining functions, called *convolution*. The convolution of two functions f and g defined on the whole real line is another function denoted $f * g$, and given by

$$f * g(x) = \int_{-\infty}^{\infty} f(t)g(x - t)\, dt. \qquad (4\text{-}39)$$

Thus, for example, the function v in (4-16) is the convolution of g and the function $P_y\colon P_y(x) = y/\pi(x^2 + y^2)$. The function u in (1-10) is an example of a slightly different convolution, appropriate to periodic functions; see Exercise 4-31.

The infinite range of integration in (4-39) raises the question of convergence of the integral. In this section we give some simple conditions for the convergence of the integral (4-39), and show that under appropriate conditions

$$\widehat{f * g} = 2\pi \hat{f} \hat{g}$$

and

$$\widehat{fg} = \hat{f} * \hat{g}.$$

Before beginning the proofs, notice that if the integrals implied by the convolution signs all exist, then convolution is a linear operation in the sense that

$$(f + h) * g = f * g + h * g$$
$$f * (g + h) = f * g + f * h, \tag{4-40}$$

and commutative, that is,

$$f * g = g * f. \tag{4-41}$$

The last relation comes from the substitution $t = x - s$ in the definition (4-39).

Theorem 4-8. *Let f and g be continuous functions.*

(i) *If $\int_{-\infty}^{\infty} |f|$ converges and $|g| \leq M$, then the integral (4-39) converges, and $f * g$ is continuous, and*

$$|f * g| \leq M \int_{-\infty}^{\infty} |f|. \tag{4-42}$$

(ii) *If $\int_{-\infty}^{\infty} |f|^2$ and $\int_{-\infty}^{\infty} |g|^2$ converge, then (4-39) converges, $f * g$ is continuous, and*

$$|f * g|^2 \leq \left(\int_{-\infty}^{\infty} |f|^2 \right) \left(\int_{-\infty}^{\infty} |h|^2 \right).$$

(iii) *If $\int_{-\infty}^{\infty} |f|$ and $\int_{-\infty}^{\infty} |g|$ converge, and g is bounded, then $\int_{-\infty}^{\infty} |f * g|$ converges to a value $< \left(\int_{-\infty}^{\infty} |f| \right) \left(\int_{-\infty}^{\infty} |g| \right)$, and*

$$\widehat{f * g} = 2\pi \hat{f} \cdot \hat{g}. \tag{4-43}$$

(iv)   *If* $\int_{-\infty}^{\infty} |f|$, $\int_{-\infty}^{\infty} |g|$, *and* $\int_{-\infty}^{\infty} |\hat{g}|$ *all converge, and g equals its Fourier integral, then*

$$\widehat{fg} = \hat{f} * \hat{g}. \tag{4-44}$$

Proof.   Under the conditions of part (i), it is clear that the integral for $f * g$ converges uniformly, and that (4-42) holds. From the uniform convergence follows the continuity of $f * g$, in view of Theorem 4-1.

In case (ii), the uniform convergence of the integral for $f * g$ follows from Schwarz's inequality:

$$\left| \int_{|t|>A} f(t)g(x-t)\,dt \right|^2$$
$$\leq \left(\int_{|t|>A} |f(t)|^2\,dt \right) \left(\int_{-\infty}^{\infty} |g(x-t)|^2\,dt \right)$$
$$= \left(\int_{|t|>A} |f(t)|^2\,dt \right) \left(\int_{-\infty}^{\infty} |g(s)|^2\,ds \right)$$
$$\to 0 \qquad \text{as } A \to \infty, \text{ independently of } x.$$

The claims in part (ii) are thus easy to establish.

In case (iii),

$$\int_{-A}^{B} \left| \int_{-\infty}^{\infty} f(t)g(x-t)\,dt \right| dx$$
$$\leq \int_{-A}^{B} \int_{-\infty}^{\infty} |f(t)g(x-t)|\,dt\,dx$$
$$= \int_{-\infty}^{\infty} |f(t)| \int_{-A}^{B} |g(x-t)|\,dx\,dt$$
$$\leq \int_{-\infty}^{\infty} |f(t)|\,dt \int_{-\infty}^{\infty} |g(x-t)|\,dx$$
$$= \left(\int_{-\infty}^{\infty} |f| \right) \left(\int_{-\infty}^{\infty} |g| \right) \tag{4-45}$$

so $\int_{-\infty}^{\infty} |f * g|$ converges, and is bounded by the last member of (4-45).   The change of order of integration is justified by part (ii) of Theorem 4-1.   To obtain the formula for the Fourier transform, we have

$$\widehat{f * g}(\xi) = \frac{1}{2\pi} \int_{-\infty}^{\infty} e^{-i\xi x} \left[\int_{-\infty}^{\infty} f(t)g(x-t)\,dt \right] dx$$
$$= \frac{1}{2\pi} \int_{-\infty}^{\infty} \left[\int_{-\infty}^{\infty} e^{-i\xi t}f(t)e^{-i\xi(x-t)}g(x-t)\,dt \right] dx$$
$$= 2\pi \hat{f}(\xi)\hat{g}(\xi).$$

The last equality comes from a change of order of integration justified by part (iv) of Theorem 4-1.

Part (iv) has the same justification:

$$f * \hat{g}(\xi) = \int_{-\infty}^{\infty} \hat{f}(\xi - \eta)\hat{g}(\eta) \, d\eta$$

$$= \frac{1}{2\pi} \int_{-\infty}^{\infty} \int_{-\infty}^{\infty} f(x)e^{-i(\xi-\eta)x}\hat{g}(\eta) \, dx \, d\eta$$

$$= \frac{1}{2\pi} \int_{-\infty}^{\infty} \int_{-\infty}^{\infty} f(x)e^{-i\xi x}\hat{g}(\eta)e^{i\eta x} \, d\eta \, dx$$

$$= \frac{1}{2\pi} \int_{-\infty}^{\infty} f(x)g(x)e^{-i\xi x} \, dx$$

since we have assumed that g equals its Fourier integral.

The assumptions made in Theorem 4-8 are taken mainly for convenience. For instance, the condition that f and g be continuous can be replaced by other conditions guaranteeing the integrability of $f(t)g(x - t)$ over finite intervals, without affecting any of the conclusions. The condition that g equals its Fourier integral in part (iv) is superfluous: see Exercise 4-29. In the same part, it would be possible to replace the convergence of $\int_{-\infty}^{\infty} |\hat{g}|$ by the convergence of $\int_{-\infty}^{\infty} |g|^2$ and $\int_{-\infty}^{\infty} |f|^2$.

Exercises. **4-28.** Let

$$P_y(x) = \frac{1}{\pi} \cdot \frac{y}{x^2 + y^2} \qquad (y > 0) \tag{4-46}$$

and let g be piecewise continuous, and $\int_{-\infty}^{\infty} |g|$ converge. Then for every x

$$2P_y * g(x) \to g(x+) + g(x-).$$

$P_y * g$ is bounded, continuous, and has bounded continuous derivatives $(P_y * g)^{(n)}$ for every n, and

$$\int |(P_y * g)^{(n)}(x)| \, dx$$

converges.

4-29. If g is continuous, and $\int_{-\infty}^{\infty} |g|$ and $\int_{-\infty}^{\infty} |\hat{g}|$ converge, then

$$g(x) = \int_{-\infty}^{\infty} \hat{g}(\xi) e^{i\xi x} \, d\xi.$$

(Hint: Consider $P_y * g$ as in the previous exercise, and apply part (iii) of Theorem 4-8.)

4-30. Let f be piecewise continuous, let $\int_{-\infty}^{\infty} |f|$ and $\int_{-\infty}^{\infty} |f|^2$ be convergent, and take P_y as in (4-46). Show

    (i) $\int_{-\infty}^{\infty} |P_y * f| \le \int_{-\infty}^{\infty} |f|.$

    (ii) $(\widehat{P_y * f})(\xi) = e^{-|\xi| y} \hat{f}(\xi).$

    (iii) $P_y * f(x) = \int_{-\infty}^{\infty} e^{-|\xi| y + i\xi x} \hat{f}(\xi) \, d\xi.$

    (iv) $\int_{-\infty}^{\infty} |P_y * f|^2$ converges.

    (v) $\int_{-\infty}^{\infty} |P_y * f - f|^2 \to 0$ as $y \to 0.$

(Hint: Use Theorems 4-5 and 4-8, and Exercises 4-24 and 4-28.)

4-31. If f and g are continuous functions of period 2π, define

$$f * g(x) = \int_{-\pi}^{\pi} f(t) g(x - t) \, dt.$$

Prove that $f * g$ is continuous and periodic, and express the Fourier coefficients of $f * g$ in terms of those of f and g.

4-32. If f and g have period 2π, find the Fourier coefficients of fg in terms of those of f and g.

4-33. The solution u of Eq. (4-35), given by (4-36), can also be represented as $u = -P * g$ with

$$2\pi \hat{P}(\xi) = (1 + \xi^2)^{-1}.$$

    (i)   Find P by checking the Fourier transform of $e^{-a|x|}$.

    (ii)   If g is continuous and $\int_{-\infty}^{\infty} |g|$ converges, then u is continuous and $\int_{-\infty}^{\infty} |u|$ converges.

    (iii)   With the assumptions of part (ii), show that u' exists and equals $-P' * g$, and u' is bounded. Hence Fourier inversion holds for u.

    (iv)   Deduce directly from the representation $u = -P * g$ that $u'' - u = g$.

4-8. THE TIME-DEPENDENT HEAT EQUATION

The equation of steady state temperature distribution in plane regions has served as a motivating example for the development of Fourier series and integrals. Now we apply these results to the equation of time-dependent temperature distribution in a rod,

$$\frac{\partial^2 u}{\partial x^2} = \frac{\partial u}{\partial t} \quad (-\infty < x < \infty, \quad t > 0), \quad (4\text{-}47)$$

where $u(x, t)$ is the temperature at the point x of the rod, at time t. For simplicity, the physical constants of density, specific heat, and conductivity have been left out, since they do not affect the nature of the solution. In addition to (4-47), we assume the initial temperature distribution is known:

$$u(x, 0) = f(x) \quad (-\infty < x < \infty). \quad (4\text{-}48)$$

Now that the general behavior of Fourier transforms has been explored, we can bypass the separation-of-variables procedure and reduce Eq. (4-47) to an ordinary differential equation by taking the Fourier transform of both sides for each individual time t. We perform the calculations first without considering the hypotheses they may require, and then formulate sufficient conditions for the validity of the result in Theorem 4-19. Let

$$\hat{u}(\xi, t) = \frac{1}{2\pi} \int_{-\infty}^{\infty} e^{-i\xi x} u(x, t)\, dx.$$

Then, assuming differentiation under the integral is valid, the transform of the right side of (4-47) is

$$\frac{1}{2\pi} \int_{-\infty}^{\infty} e^{-i\xi x} \frac{\partial u(x, t)}{\partial t}\, dx$$

$$= \frac{\partial}{\partial t} \frac{1}{2\pi} \int_{-\infty}^{\infty} e^{-i\xi x} u(x, t)\, dx$$

$$= \frac{\partial \hat{u}(\xi, t)}{\partial t}.$$

According to Theorem 4-7 the transform of the left side is $-\xi^2 \hat{u}(\xi, t)$, so (4-47) becomes

$$-\xi^2 \hat{u}(\xi, t) = \frac{\partial \hat{u}(\xi, t)}{\partial t} \quad (-\infty < \xi < \infty)$$

and the initial condition (4-48) becomes

$$\hat{u}(\xi, 0) = \hat{f}(\xi).$$

These are solved by

$$\hat{u}(\xi, t) = \exp(-\xi^2 t)\hat{f}(\xi).$$

According to Theorem 4-8 on convolutions, this corresponds to

$$u(x, t) = G_t * f(x) \tag{4-49}$$

when

$$\hat{G}_t(\xi) = \frac{1}{2\pi} \exp(-\xi^2 t).$$

By the inversion formula, G_t is given by

$$G_t(x) = \frac{1}{2\pi} \int_{-\infty}^{\infty} e^{i\xi x} \exp(-\xi^2 t) \, d\xi$$

$$= \frac{1}{2\pi} \int_{0}^{\infty} \cos \xi x \exp(-\xi^2 t) \, d\xi$$

$$= \frac{1}{2\sqrt{\pi t}} \exp\left(\frac{-x^2}{4t}\right).$$

from Exercise 4-21. Hence we obtain an integral for u, comparable with the Poisson integral:

$$u(x, t) = \frac{1}{2\sqrt{\pi t}} \int_{-\infty}^{\infty} \exp\left(\frac{-(x-y)^2}{4t}\right) f(y) \, dy. \tag{4-50}$$

Theorem 4-9. *Let f be continuous, $\int_{-\infty}^{\infty} f$ converge, and u be given by (4-50). Then*

(i) $\int_{-\infty}^{\infty} |u(x, t)| \, dx$ *converges, and is bounded by $\int_{-\infty}^{\infty} |f|$.*

(ii) $\int_{-\infty}^{\infty} u(x, t) \, dx = \int_{-\infty}^{\infty} f$.

(iii)   *u and all its derivatives are continuous for $t > 0$; $\partial^2 u/\partial x^2$ = $\partial u/\partial t$; and $\int_{-\infty}^{\infty} |\partial u/\partial x|$, $\int_{-\infty}^{\infty} |\partial^2 u/\partial x^2|$, $\int_{-\infty}^{\infty} |\partial u/\partial t|$ all converge uniformly in $t \geq \delta$, for each $\delta > 0$.*

(iv)   *$u(x,t) \to f(x)$ as $t \to 0$, uniformly on every finite interval.*

(v) $\int_{-\infty}^{\infty} |u(x, t) - f(x)| \, dx \to 0$.

Conversely, if v satisfies parts (i), (iii), and (v), then $v = u$.

Proof. We leave the proof of the converse to the reader; it amounts largely to checking that the deduction made above is valid when parts (i), (iii), and (v) hold.

The claim (i) follows from Theorem 4-8(iii). Equation (ii) says, physically, that the amount of heat in the rod does not change with time. Its proof is simple:

$$\int_{-\infty}^{\infty} u(x, t)\, dx = \int_{-\infty}^{\infty} \int_{-\infty}^{\infty} G_t(x - y) f(y)\, dy\, dx$$

$$= \int_{-\infty}^{\infty} \left[\int_{-\infty}^{\infty} G_t(x - y)\, dx \right] f(y)\, dy$$

$$= \int_{-\infty}^{\infty} f(y)\, dy$$

since

$$\int_{-\infty}^{\infty} G_t(x - y)\, dx = \int_{-\infty}^{\infty} G_t(x)\, dx = 2\pi \hat{G}_t(0) = 1.$$

The differential equation in part (iii) is verified by differentiating under the integral, and the uniform convergence of the various integrals is obtained as follows. We have

$$\int_{|x| \geq A} \left| \frac{\partial u(x, t)}{\partial t} \right| dx$$

$$\leq \int_{|x| \geq A} \int_{|y| \leq B} \left| \frac{\partial G_t(x - y)}{\partial t} \right| \cdot |f(y)|\, dy\, dx$$

$$+ \int_{|x| \geq A} \int_{|y| > B} \left| \frac{\partial G_t(x - y)}{\partial t} \right| \cdot |f(y)|\, dy\, dx$$

$$\leq \int_{|y| < B} \left[\int_{|x| \geq A} \left| \frac{\partial G_t(x - y)}{\partial t} \right| dx \right] |f(y)|\, dy$$

$$+ \int_{|y| > B} \left[\int_{-\infty}^{\infty} \left| \frac{\partial G_t(x - y)}{\partial t} \right| dx \right] |f(y)|\, dy.$$

Now, given $\epsilon > 0$, the last repeated integral can be made $< \epsilon/2$ by choosing B sufficiently large. Then in the next-to-last repeated integral we have, for $|y| < B$, that

$$\int_{|x| \geq A} \left| \frac{\partial G_t(x - y)}{\partial t} \right| dx$$

$$\leq \int_{|z| \geq A - B} \left| \frac{\partial G_t(z)}{\partial t} \right| dz$$

so that the repeated integral in question is dominated by

$$\left(\int_{|y|<B} |f(y)|\,dy \right) \int_{|z|\geq A-B} \left| \frac{\partial G_t(z)}{\partial t} \right| dz$$

which is $< \epsilon/2$ for A sufficiently large, for all $t \geq \delta$.

For the convergence in part (v), we leave it to the reader to verify that, given any $\epsilon > 0$, f can be written as a sum of two continuous functions,

$$f = f_1 + f_2$$

with

$$f_1(x) = 0 \qquad \text{for} \quad |x| > B$$

and

$$\int_{-\infty}^{\infty} |f_2| < \frac{\epsilon}{4}.$$

Granted this, we have

$$\int_{-\infty}^{\infty} |u - f| \leq \int_{-\infty}^{\infty} |G_t * f_1(x) - f_1(x)|\,dx$$
$$+ \int_{-\infty}^{\infty} |G_t * f_2| + \int_{-\infty}^{\infty} |f_2|. \quad (4\text{-}51)$$

Since $\int_{-\infty}^{\infty} |G_t| = 1$, the last two terms add up to $< \epsilon/2$. The first term on the right of (4-51) is split into an integral for $|x| < 2B$, and another for $|x| > 2B$. The integral for $|x| < 2B$ is $< \epsilon/4$ for t sufficiently small, because of the uniform convergence claimed in part (iv). For the final term we have, since f_1 vanishes for $|x| > B$, and $G_t(z)$ is positive, even, and monotone decreasing in $z \geq 0$, that

$$\int_{|x|>2B} |G_t * f_1(x) - f_1(x)|\,dx$$
$$\leq \int_{|x|>2B} \int_{-B}^{B} |G_t(x-y)f_1(y)|\,dy\,dx$$
$$= \int_{-B}^{B} \left[\int_{|x|>2B} |G_t(x-y)|\,dx \right] |f_1(y)|\,dy$$
$$\leq 2 \int_{-B}^{B} \left[\int_{2B}^{\infty} G_t(x-B)\,dx \right] |f_1(y)|\,dy.$$

Here again

$$\int_{2B}^{\infty} G_t(x-B)\,dx = \int_{B}^{\infty} G_t(x)\,dx$$

tends to zero as $t \to 0$, so finally the terms on the right of (4-51) add up to $< \epsilon$ if t is sufficiently small.

This concludes Theorem 4-19. The exercises below bring out further details of the solution of this temperature problem.

Exercises. 4-34. Suppose the initial temperature f in (4-50) is real-valued. Prove that

$$\inf f \leq u \leq \sup f.$$

4-35. Prove the analog of Theorem 4-19 for the equation

$$\frac{\partial^2 u}{\partial x^2} + \frac{\partial^2 u}{\partial y^2} = 0$$

with the kernel G_t in (4-49) replaced by the Poisson kernel P_y in (4-46).

4-36. The problem of time-dependent temperature distribution in a circular loop of length 2π is described by Eqs. (4-47) and (4-48), together with the condition that f and u have period 2π in the variable x. Suppose u is continuous for $t \geq 0$ and $u(x, 0) = f(x)$. Verify the following for the solution u.

(i) $\quad u(x, t) = \displaystyle\int_{-\pi}^{\pi} \sum_{-\infty}^{\infty} G_t(x + 2\pi n - y)f(y) \, dy, \ t > 0.$

(ii) $\quad u(x, t) = \displaystyle\sum_{-\infty}^{\infty} a_n e^{inx} \exp(-n^2 t), \ t > 0,$

with a_n the Fourier coefficients of f.

(iii) $\quad \inf f \leq u \leq \sup f$ if f is real-valued.

(iv) $\quad \lim_{t \to \infty} u(x, t) = (1/2\pi) \displaystyle\int_{-\pi}^{\pi} f(y) \, dy.$

4-37. Use an appropriate symmetry to reduce the problem of heat conduction in a rod of length π with insulated ends to the problem of the previous exercise. Obtain the results analogous to parts (i)–(iv) of that problem.

Bibliography

This list is just a sample of the extensive literature on Fourier series and transforms. References [12] and [13] are historical surveys; [5], [15], and [10] are standard mathematical introductions, the latter based on the Lebesgue integral; [6], [7], [9], and [11] concern applications more or less within the context of this book. References [1] and [3] consider more sophisticated applications to optics and probability; they require Lebesgue integration. Reference [17] is a thorough introduction to the Lebesgue integral, Fourier series, and generalizations. References [2], [4], [18], and [19] are general theoretical treatises.

1. J. Arsac, *Transformation de Fourier et Théorie des Distributions*, Dunod, Paris, 1961.
2. N. Bary, *A Treatise on Trigonometric Series*, 2 vols., M. F. Mullins, transl., Macmillan, New York, 1964.
3. S. Bochner, *Harmonic Analysis and the Theory of Probability*, University of California Press, Berkeley, California, 1955.
4. S. Bochner and K. Chandrasekharan, *Fourier Transforms*, Princeton University Press, Princeton, New Jersey, 1949.

5.   H. Carslaw, *Theory of Fourier's Series and Integrals*, 3d ed., Macmillan, London, and Dover, New York, 1930.

6.   H. S. Carslaw and J. C. Jaeger, *Conduction of Heat in Solids*, 2d ed., Oxford University Press, London, 1959.

7.   R. V. Churchill, *Fourier Series and Boundary Value Problems*, 2d ed., McGraw-Hill, New York, 1963.

8.   A. Erdélyi *et al.*, *Tables of Integral Transforms*, Vol. 1, McGraw-Hill, New York, 1954.

9.   P. Franklin, *Fourier Methods and the Laplace Transform*, Dover, New York, 1958.

10. G. H. Hardy and W. W. Rogosinski, *Fourier Series*, University Press, Cambridge, 1956.

11. L. Hopf, *Differential Equations of Physics*, Walter Nef, transl., Dover, New York, 1948.

12. R. L. Jeffery, *Trigonometric Series*, University of Toronto Press, Toronto, 1956.

13. R. E. Langer, *Fourier's Series, The Genesis and Evolution of a Theory*, Supplement to the *American Mathematical Monthly*, 1947.

14. F. Oberhettinger, *Tabellen zur Fourier Transformationen*, Springer, Berlin, 1957.

15. W. Rogosinski, *Fourier Series*, H. Cohn and F. Steinhardt, transl., Chelsea, New York, 1950.

16. A. Sommerfeld, *Partial Differential Equations in Physics*, E. G. Straus, transl., Academic Press, New York, 1949.

17. B. Sz. Nagy, *Introduction to Real Functions and Orthogonal Expansions*, Oxford University Press, New York, 1965.

18. E. C. Titchmarsh, *Theory of Fourier Integrals*, Oxford University Press, London, 1937.

19. A. Zygmund, *Trigonometric Series*, 1st ed. reprinted by Dover, New York, 1955; 2nd ed., 2 vols., University Press, Cambridge, 1959.

Answers to Exercises

1-1. $u_{xx} + u_{yy} = 0$.

1-2. $u = \log r$ satisfies (1-2) and (1-4); $u = r^{1/2} \cos \theta/2$ satisfies (1-2) and (1-3).

1-3. $\sin^2 \theta = \frac{1}{2} - \dfrac{\cos 2\theta}{2} = -\frac{1}{4}e^{-i2\theta} + \frac{1}{2} - \frac{1}{4}e^{i2\theta}$

$\cos^3 \theta = \frac{3}{4} \cos \theta + \frac{1}{4} \cos 3\theta$

$\qquad = \frac{1}{8} e^{-i3\theta} + \frac{3}{8} e^{-i\theta} + \frac{3}{8} e^{i\theta} + \frac{1}{8} e^{i3\theta}.$

1-4. $\dfrac{1}{2\pi} \displaystyle\int_{-\pi}^{\pi} |\theta| \, e^{in\theta} \, d\theta = \dfrac{1}{2\pi} \int_{0}^{\pi} \theta (e^{in\theta} + e^{-in\theta}) \, d\theta$

$\qquad\qquad\qquad = \dfrac{1}{\pi} \displaystyle\int_{0}^{\pi} \theta \cos n\theta \, d\theta = \dfrac{\pi}{2} \quad$ if $\quad n = 0$

$\qquad\qquad\qquad\qquad = 0 \quad$ if $\quad n$ is even and nonzero

$\qquad\qquad\qquad\qquad = \dfrac{-2}{\pi n^2} \quad$ if $\quad n$ is odd

$u(r, \theta) = \dfrac{\pi}{2} - 2 \displaystyle\sum_{-\infty}^{\infty} \dfrac{r^{|2k-1|} \, e^{i(2k-1)\theta}}{\pi(2k-1)^2}$

$\qquad\quad = \dfrac{\pi}{2} - 4 \displaystyle\sum_{1}^{\infty} \dfrac{r^{2k-1} \, [\cos(2k-1)\theta]}{\pi \, (2k-1)^2}$

When $r \leq 1$, $\left| u(r, \theta) - \pi/2 + 4\Sigma_1^K r^{2k-1}[\cos(2k-1)\theta]/\pi(2k-1)^2 \right|$ < 0.1 for any $K > 3$, by integral test.

$$0 \leq u(\tfrac{1}{2}, \pi) - \frac{\pi}{2} - \frac{37}{18\pi} < 0.005.$$

1-8. $r^n e^{in\theta} = (re^{i\theta})^n = (x + iy)^n;\; r^n e^{-in\theta} = (x - iy)^n.$

1-14. $\pi^2/8 = \Sigma_1^\infty (2k-1)^{-2}.$

1-20. Draw graphs of t and $\pi \sin(t/2)$.

1-21. (a) $\theta = -2\Sigma_1^\infty (-1)^n n^{-1} \sin n\theta,\; -\pi < \theta < \pi.$

(b) $\theta^2 = (\pi^2/3) + 4\Sigma_1^\infty n^{-2}(-1)^n \cos n\theta,\; -\pi \leq \theta \leq \pi.$

(c) $|\theta| = (\pi/2) - (4/\pi)\Sigma_1^\infty (2k-1)^{-2} \cos(2k-1)\theta,$
$-\pi \leq \theta \leq \pi.$

(d) $1 = (4/\pi)\Sigma_1^\infty (2k-1)^{-1} \sin(2k-1)\theta,\; 0 < \theta < \pi.$

Notice that term-by-term differentiation of (b) yields (a), and of (c) yields (d).   See Exercise 1-24..

1-22. (a)
$$\frac{\pi}{4} = \sum_1^\infty \frac{(-1)^{k+1}}{(2k-1)}$$

$$\frac{\pi}{3\sqrt{3}} = 1 - \tfrac{1}{2} + \tfrac{1}{4} - \tfrac{1}{5} + \tfrac{1}{7} - \tfrac{1}{8} + \tfrac{1}{10} - \cdots$$

$$\frac{\pi}{8} = \frac{1}{\sqrt{2}} (1 + \tfrac{1}{3} - \tfrac{1}{5} - \tfrac{1}{7} + \tfrac{1}{9} + \cdots)$$
$$+ (-\tfrac{1}{2} + \tfrac{1}{6} - \tfrac{1}{10} + \tfrac{1}{14} - \cdots).$$

(b)
$$\frac{\pi^2}{6} = \sum_1^\infty n^{-2}$$

$$\frac{\pi^2}{12} = 1 - 2^{-2} + 3^{-2} - 4^{-2} + \cdots$$

$$\frac{\pi^2}{9} = (1 + 2^{-2} - 2 \cdot 3^{-2} + 4^{-2} + 5^{-2} - 2 \cdot 6^{-2} + 7^{-2}$$
$$+ \cdots).$$

(c)
$$\frac{\pi^2}{8} = \sum_1^\infty (2k-1)^{-2}$$

$$\frac{\pi^2}{12} = (1 - 2 \cdot 3^{-2} + 5^{-2} + 7^{-2} - 2 \cdot 9^{-2} + 11^{-2}$$
$$+ \cdots)$$

$$\frac{\pi^2}{8\sqrt{2}} = (1 - 3^{-2} - 5^{-2} + 7^{-2} + 9^{-2} - 11^{-2} - 13^{-2}$$
$$+ \cdots).$$

(d) $\frac{\pi}{4} = \sum_{1}^{\infty} (-1)^{k+1}/(2k-1)$

$= \frac{\sqrt{3}}{2} (1 - \frac{1}{5} + \frac{1}{7} - \frac{1}{11} + \frac{1}{13} - \frac{1}{17} + \cdots)$

$= \frac{1}{\sqrt{2}} (1 + \frac{1}{3} - \frac{1}{5} - \frac{1}{7} + \frac{1}{9} + \cdots).$

2-1. $u(x, t) = \sum_{1}^{\infty} (a_n e^{itn\omega} + b_n e^{-itn\omega}) \cos \frac{n x \pi}{L}.$

2-4. $2v\left(\frac{\pi x}{L}\right) = f(x) + \frac{\pi}{L\omega} \int_0^x g \qquad 0 \leq x \leq L$

$2v\left(-\frac{\pi x}{L}\right) = \frac{\pi}{L\omega} \int_0^x g - f(x) \qquad 0 \leq x \leq L.$

2-5.   (i) $a_n = b_n = \frac{1}{L}\left[\int_0^{L/2} \frac{x}{4} \sin \frac{n x \pi}{L} \, dx \right.$

$\left. + \int_{L/2}^{L} \left(\frac{L}{4} - \frac{x}{4}\right) \sin \frac{n x \pi}{L} \, dx \right].$

(ii) $a_n = b_n = \frac{1}{L}\left[\int_0^{L/4} \frac{x}{2} \sin \frac{n x \pi}{L} \, dx \right.$

$\left. + \int_{L/4}^{L} \left(\frac{L}{6} - \frac{x}{6}\right) \sin \frac{n x \pi}{L} \, dx \right].$

(iii)   Amplitude $= 2 |a_n|$; average kinetic energy $= n^2\omega^2\rho L a_n^2/2$. The first harmonic $(n = 2)$ is not present in case (i), but fairly strong in case (ii).

2-6.   (b) $X_n T_n = \cos \frac{n x \pi}{L} \exp\left(-\frac{tkn^2\pi^2}{c\rho L^2}\right) \qquad (n = 0, 1, \ldots).$

(d) $u(x, t) = \frac{L}{2} - \frac{4L}{\pi^2} \sum_{1}^{\infty} \frac{\cos(2j-1)\pi x/L}{(2j-1)^2} \exp\left(\frac{tk(2j-1)^2\pi^2}{-c\rho L^2}\right).$

(e) $(1/L) \int_0^L u(x, 0) \, dx.$ Yes.

(f) $(c\rho/L) \int_0^L u(x, 0) \, dx.$ Yes.

3-2. $f = 0.$

3-5. $\displaystyle\sum_{1}^{\infty} n^{-4} = \pi^4/90$.

3-7. (i) $a_0 = 0$; (ii) $a_0 = \pi^2/4$; (iii) $a_0 = \pi^2/3$; (iv) $a_0 = \pi^2/2$.

3-13. $\displaystyle\int_{-\pi}^{\pi} f\bar{g} = 2\pi \sum_{-\infty}^{\infty} a_n \bar{b}_n$.

3-14. (ii) and (iv) are absolutely convergent.

4-8. $v(x, y)$ is constant.

4-16. $w = xy$ is one solution, but there is a simpler one.

4-20. $\pi e^{-|at|} |a|$.

4-21. $c = 1/(2a \sqrt{\pi})$.

4-25.
$$\int_{-\infty}^{\infty} a^2(\xi^2 + a^2)^{-2} \, d\xi = \frac{\pi}{2a}$$

$$\int_{-\infty}^{\infty} \xi^{-2}[1 - \cos(a\xi - b\xi)] \, d\xi = \pi \, |b - a|$$

$$\int_{-\infty}^{\infty} \xi^{-4}(\xi \cos \xi - \sin \xi)^2 \, d\xi = \frac{\pi}{3}.$$

4-31. If a_n, b_n, c_n are respectively the Fourier coefficients of f, g, and $f * g$, then $c_n = 2\pi a_n b_n$.

4-32. $c_n = \Sigma_{m+k=n} a_k b_m = \Sigma_{j=-\infty}^{\infty} a_{j-m} b_m$, a sort of convolution of the sequences $\{a_n\}$ and $\{b_n\}$.

Index

CPSIA information can be obtained
at www.ICGtesting.com
Printed in the USA
LVHW081412130121
676373LV00019B/963